Fundamentals of
Glycosylation

Edited by Alok Raghav and Jamal Ahmad

Published in London, United Kingdom

IntechOpen

Supporting open minds since 2005

Fundamentals of Glycosylation
http://dx.doi.org/10.5772/intechopen.90986
Edited by Alok Raghav and Jamal Ahmad

Contributors
Akio Nakamura, Ritsuko Kawaharada, James C. Samuelson, Minyong Chen, Steven J. Dupard, Colleen M. McClung, Saulius Vainauskas, Mehul B. Ganatra, Christopher H. Taron, Cristian I. Ruse, Iryna Brodyak, Natalia Sybirna, Alok Raghav, Jamal Ahmad, Renu Tomar

Notice
Statements and opinions expressed in the chapters are these of the individual contributors and not necessarily those of the editors or publisher. No responsibility is accepted for the accuracy of information contained in the published chapters. The publisher assumes no responsibility for any damage or injury to persons or property arising out of the use of any materials, instructions, methods or ideas contained in the book.

First published in London, United Kingdom, 2022 by IntechOpen
IntechOpen is the global imprint of INTECHOPEN LIMITED, registered in England and Wales, registration number: 11086078, 5 Princes Gate Court, London, SW7 2QJ, United Kingdom
Printed in Croatia

British Library Cataloguing-in-Publication Data
A catalogue record for this book is available from the British Library

Additional hard and PDF copies can be obtained from orders@intechopen.com

Fundamentals of Glycosylation
Edited by Alok Raghav and Jamal Ahmad
p. cm.
Print ISBN 978-1-83969-134-8
Online ISBN 978-1-83969-135-5
eBook (PDF) ISBN 978-1-83969-136-2

We are IntechOpen,
the world's leading publisher of
Open Access books
Built by scientists, for scientists

5,600+
Open access books available

138,000+
International authors and editors

175M+
Downloads

Our authors are among the

156
Countries delivered to

Top 1%
most cited scientists

12.2%
Contributors from top 500 universities

Interested in publishing with us?
Contact book.department@intechopen.com

Numbers displayed above are based on latest data collected.
For more information visit www.intechopen.com

Meet the editors

Dr. Alok Raghav pursued his doctoral degree in Endocrinology from Rajiv Gandhi Centre for Diabetes and Endocrinology, J.N. Medical College, Faculty of Medicine, Aligarh Muslim University, Aligarh, India. He worked as a project scientist at the Indian Institute of Technology Kanpur, India. He has more than ten years of research experience in the field of glycobiology and diabetes mellitus. He is currently a Scientist C at the Multidisciplinary Research Unit (sponsored by Department of Health Research, Ministry of Health and Family Welfare, New Delhi), GSVM Medical College Kanpur, India. He is also an associate editor and editor for several peer-reviewed journals. Dr. Raghav has received several international and national awards.

Dr. Jamal Ahmad pursued his doctoral degree in Medicine at Aligarh Muslim University, Aligarh, India. He obtained his D.Sc degree in Endocrinology from the Faculty of Medicine, Aligarh Muslim University, Aligarh. He served as Director Rajiv Gandhi Centre for Diabetes and Endocrinology, J.N. Medical College, Faculty of Medicine, Aligarh Muslim University. He also held the position of Dean, Faculty of Medicine, Aligarh Muslim University, Aligarh. He has over 40 years of research experience in the field of glycobiology, medicine, endocrinology, and diabetes mellitus. He is serving as associate editor and editor of several peer-reviewed journals and recipient of several International and National Awards. He is also serving as an expert member of various government constituted committees.

Contents

Preface

This book describes the fundamental concept of glycosylation and reflects the wide range of research currently being practiced in the field of proteins. Apart from proteins, most of the concepts of glycosylation are applicable to other biological macromolecules including nucleic acids and lipids. Chapter 2 provides an overview of advanced glycation end products to explain non-enzymatic glycation.

Since the emergence of newer techniques like proteomics and lipidomics, deciphering the site of glycosylation within proteins and lipids respectively has now enabled us to identify the targeted site upon which the glycosylation occurred. To give researchers and non-specialist readers a scientific understanding, this book explains not only the concept of glycosylation but also discusses recent techniques of glycoproteomics to screen the sites of glycosylation. Numerous approaches of performing glycosylation enrichment, which has great importance in medicine and health, are also explained.

Sialylation, similar to glycosylation, also has a role in modifying biological macromolecules, including various intracellular and extracellular proteins. The overview and understanding of sialic acid, glycans, and glycoconjugates simultaneously with the mechanism of glycosylation provide additional insights.

Since the introduction of SARS-CoV-2, several new variants of coronaviruses have been identified. SARS-CoV-2 is a potential threat to the human community, and it is very important to know every aspect of this virus infection, phage transmission, and its regulation within the cells. Therefore, in this edition of the book, we include a chapter describing the role of post-translational modification that has a significant role in SARS-CoV-2 infection. The book also discusses another post-translational modification including palmitoylation and phosphorylation of the viral spike and envelope proteins along with their significance in the pathogenesis of the viral infection phase. Because the current understanding and knowledge for SARS-CoV-2 is limited, it is very important for readers to understand its pathogenesis mechanism and post-translational modifications.

We would like to thank the contributing authors for the time they have given in preparing their chapters and providing essential knowledge and scientific understanding of concepts of glycosylation. We would also like to acknowledge the Author Service Manager at IntechOpen for assisting at every step of the book publication.

Alok Raghav
Multidisciplinary Research Unit,
GSVM Medical College,
Kanpur, India

Jamal Ahmad
J.N Medical College,
Aligarh Muslim University,
Aligarh, India

Section 1

Introduction

Chapter 1

Introductory Chapter: Glimpses of Glycosylation

Alok Raghav and Jamal Ahmad

1. Introduction

Glycosylation refers to the post translational modification of proteins, lipids and nucleic acids in the presence of enzymes. The common post translational modifications of biological importance include N-linked glycosylation, O-linked glycosylation, phospho–serine glycosylation, as well as C-mannosylation and glypation (addition of glycophosphatidylinositol. Glycosylation reaction is mediated by glycosyltransferases that conjugate carbohydrate to proteins, lipids and nucleic acids. Glycosylation is a biological mechanism that aids the protein folding and help in protein signaling along with cell-cell interaction. Recently mass spectrometry technique is recently gained attention in quantitative estimation of glycoproteomics. In COVID-19 infection, the SARS-CoV-2 viral proteins including spike(S), envelope (C), membrane (M) and nucleocapsid (N) undergo post translational modification of glycosylation and phosphorylation that plays important role in virulence and infectivity.

Defined in the broadcast sense, glycosylation is the conjugation process of joining carbohydrate to the protein's backbone via enzymatic reaction. Post modification after this reaction, the protein is termed as glycoprotein. In our body, the most common glycosylation reaction refers to the N-glycosylation and O-glycosylation. Among these two post modification process, the N-glycosylation is frequently occurring mechanism. This mechanism of post translation modification comes under the domain of glycobiology, which refers to the study of biosynthesis, structures and biology of saccharides (also known as sugar or carbohydrates). Glycosylation is a critical mechanism of the biosynthetic-secretory pathway that occurs in endoplasmic reticulum (ER) and Golgi apparatus [1]. I was a known fact that nearly half of the eukaryotic proteins undergo modification, which presents the covalent conjugation of sugar moieties to particular amino acids of interest [2]. Membrane-bound and soluble proteins along with secreted proteins, ligands, surface proteins that are transported from the Golgi to the cytoplasm also become glycosylated [2].

Lipids and proteoglycans are also among the susceptible targets of the glycosylation that significantly contribute in increasing the number of substrate for the post translation modification. Carbohydrate plays an important role in the regulation of multicellular organs and organism's functions as a resultant several biological macromolecules possess covalently attached saccharides including monosaccharides and oligosaccharides which are commonly known as glycans [3]. These glycans contribute a major role in modulating cell-to-cell interaction, development and function of cell, interaction between cell and host [4]. Protein bound glycans are abundant in the nucleus and cytoplasm and serve as the key regulator element there [4].

2. N- and O-glycosylation

Several proteins undergoes post translational modification by N-glycosylation that denote with the attachment of the N-acetylglucosamine (GlcNAc) to the nitrogen atom present in the Asn side chain of the amino acid by a β-1 N linkage [2]. These GlcNAc2 mannose (Man)3 core containing glycoconjugates shows the tendency to add/remove several monosaccharides. These conjugation reactions include galactosylation, sialylation, GlcNAclyation and fucosylation [2]. Moreover, glycosylation occur on hydroxyl functional group of the amino acids Ser and Thr. The most common sugars showing conjugation with Ser and Thr are GlcNAc and N-acetylgalactosamine (GalNAc) [5]. GalNAc associated glycans, often known as mucin-type O-glycans are abundantly found in extracellular spaces and secreted proteins like mucin [6]. Mucin is characterized with high number of Pro, Ser and Thr residues that makes it susceptible to the O-linked glycosylation. The participating sugars conjugate with the protein as it moves through the cis, medial and trans Golgi apparatus. The glycopeptide O-glycans are post transnationally modified by glycosyltransferases [5].

3. C-Mannosylation

C-mannosylation is different approach from glycosylation, as in this process there is reaction forms carbon-carbon bonds rather than carbon-nitrogen and carbon-oxygen bonds. C-mannosyltransferase is the key enzyme that is involved in the C-mannosylation. This enzyme conjugates the C1 of the mannose to C2 of the indole ring present in the tryptophan residue [7]. Moreover, this enzyme recognizes the specific sequence of Trp-X-X-Trp that further ease the transfer of mannose sugar from dolichol-P-Man to the first Trp residue present in the given sequence [8–10]. In another event of the post translational modification (PTM), the phosphoglycosylation is among the PTM mechanism that is limited to the parasites including Trypanosoma and Leishmania including slime molds. The mechanism is characterized by the conjugation of glycans to the Ser and Thr residues linked by the phosphodiester bonds [11].

4. Glycosylation vs. glycation

Glycosylation and glycation are enzymatic and non-enzymatic reaction respectively, in which glucose and other glucose metabolites show affinity with different biological macromolecules such as proteins, lipids and nucleic acids. Increased presence of glucose both intracellular and extracellular possesses threat to the human and act as risk factor for various metabolic disorders including diabetes mellitus. Both non-enzymatic glycation and enzymatic glycosylation reaction have been shown to play important role in human health. Non-enzymatic glycation leads to the formation of advanced glycation end products (AGEs) that generates as a resultant of protein and lipid glycation with aldose sugars [12, 13]. Prior to the formation of AGEs, the early glycation reaction occurs that form Schiff bases and Amadori products. AGEs formed due to the non-enzymatic glycation generates reactive oxygen species (ROS), that further binds to the particular cell surface receptors and later forms cross link [12, 14].

5. SARS-CoV-2 and glycosylation

Severe acute respiratory syndrome coronavirus 2 (SARS-CoV2) is severely affecting the worldwide population. SARS-CoV-2 virulence and survival is impacted by the glycans that is covalently linked to the proteins of virus through the process of glycosylation that makes alteration in the virulence of the SARS-CoV-2 virus. It belongs to the coronavirus family which exhibit protein constituted enveloped single-stranded RNA. These viral proteins undergo post-translational modifications (PTMs) that reorganized covalent bonds and modify the polypeptides and in turn modulate the protein functions. Being viral machinery, it uses host cells system to replicate itself and make their copes, their proteins are also subject to PTMs. Glycosylation, palmitoylation of the spike and envelope proteins, phosphorylation, of the nucleocapsid protein are among the major PTMs responsible for the pathogenesis of the viral infection phase. The current knowledge of CoV proteins PTMs is limited and need to be exploring for to understand the viral pathogenesis mechanism and PTMs effect of infection phase.

Acknowledgements

The author Dr. Alok Raghav is thankful to the Department of Health Research, Ministry of Health and Family Welfare, New Delhi for providing financial support in the form of salary.

Conflict of interest

The authors declare no conflict of interest.

Notes/thanks/other declarations

Place any other declarations, such as "Notes", "Thanks", etc. in before the References section. Assign the appropriate heading. Do NOT put your short biography in this section. It will be removed.

Author details

Alok Raghav[1*] and Jamal Ahmad[2]

1 Multidisciplinary Research Unit, Ganesh Shankar Vidyarthi Memorial Medical College, Kanpur, India

2 J.N. Medical College, Aligarh Muslim University, Aligarh, India

*Address all correspondence to: alokalig@gmail.com

IntechOpen

References

[1] Zhang X. Alterations of golgi structural proteins and glycosylation defects in cancer. Frontiers in Cell and Developmental Biology. 2021;**9**:665289. DOI: 10.3389/fcell.2021.665289

[2] Reily C, Stewart TJ, Renfrow MB, Novak J. Glycosylation in health and disease. Nature Reviews Nephrology. 2019;**15**:346-366. DOI: 10.1038/s41581-019-0129-4

[3] Smith DF, Song X, Cummings RD. Use of glycan microarrays to explore specificity of glycan-binding proteins. Methods in Enzymology. 2010;**480**:417-444. DOI: 10.1016/S0076-6879(10)80033-3

[4] Varki A, Gagneux P. Biological functions of glycans. In: Varki A, Cummings RD, Esko JD, et al., editors. Essentials of Glycobiology [Internet]. 3rd ed. Cold Spring Harbor (NY): Cold Spring Harbor Laboratory Press; 2015-2017. Chapter 7. DOI: 10.1101/glycobiology.3e.007

[5] Varki A, Cummings RD, Esko JD, Stanley P, Hart GW, Aebi M, Darvill AG, Kinoshita T, Packer NH, Prestegard JH, Schnaar RL, Seeberger PH, editors. Essentials of Glycobiology [Internet]. 3rd ed. Cold Spring Harbor (NY): Cold Spring Harbor Laboratory Press; 2015-2017

[6] Vasudevan D, Haltiwanger RS. Novel roles for O-linked glycans in protein folding. Glycoconjugate Journal. 2014; **31**:417-426

[7] de Beer T, Vliegenthart JF, Löffler A, Hofsteenge J. The hexopyranosyl residue that is C-glycosidically linked to the side chain of tryptophan-7 in human RNase Us is alpha-mannopyranose. Biochemistry. 1995;**34**(37):11785-11789. DOI: 10.1021/bi00037a016

[8] Krieg J, Hartmann S, Vicentini A, Gläsner W, Hess D, Hofsteenge J. Recognition signal for C-mannosylation of Trp-7 in RNase 2 consists of sequence Trp-x-x-Trp. Molecular Biology of the Cell. 1998;**9**(2):301-309. DOI: 10.1091/mbc.9.2.301

[9] Doucey MA, Hess D, Cacan R, Hofsteenge J. Protein C-mannosylation is enzyme-catalysed and uses dolichyl-phosphate-mannose as a precursor. Molecular Biology of the Cell. 1998;**9**(2):291-300. DOI: 10.1091/mbc.9.2.291

[10] Hartmann S, Hofsteenge J. Properdin, the positive regulator of complement, is highly C-mannosylated. Journal of Biological Chemistry. 2000;**275**(37):28569-28574. DOI: 10.1074/jbc.M001732200

[11] Haynes PA. Phosphoglycosylation: A new structural class of glycosylation? Glycobiology. 1998;**8**(1):1-5. DOI: 10.1093/glycob/8.1.1

[12] Schmidt AM, Hori O, Brett J, Yan SD, Wautier JL, Stern D. Cellular receptors for advanced glycation end products: Implications for induction of oxidant stress and cellular dysfunction in the pathogenesis of vascular lesions. Arteriosclerosis, Thrombosis, and Vascular Biology. 1994;**14**:1521-1528

[13] Singh R, Barden A, Mori T, Beilin L. Advanced glycation end-products: A review. Diabetologia. 2001;**44**:129-146

[14] Brownlee M, Vlassara H, Cerami A. Nonenzymatic glycosylation products on collagen covalently trap low-density lipoprotein. Diabetes. 1985;**34**:938-941

Section 2

Advanced Glycation
End Products

Advanced Glycation End Products and Oxidative Stress in a Hyperglycaemic Environment

Akio Nakamura and Ritsuko Kawaharada

Abstract

Protein glycation is the random, nonenzymatic reaction of sugar and protein induced by diabetes and ageing; this process is quite different from glycosylation mediated by the enzymatic reactions catalysed by glycosyltransferases. Schiff bases form advanced glycation end products (AGEs) via intermediates, such as Amadori compounds. Although these AGEs form various molecular species, only a few of their structures have been determined. AGEs bind to different AGE receptors on the cell membrane and transmit signals to the cell. Signal transduction via the receptor of AGEs produces reactive oxygen species in cells, and oxidative stress is responsible for the onset of diabetic complications. This chapter introduces the molecular mechanisms of disease onset due to oxidative stress, including reactive oxygen species, caused by AGEs generated by protein glycation in a hyperglycaemic environment.

Keywords: glycation, advanced glycation end products, gestational diabetes, reactive oxygen species, oxidative stress

1. Introduction

Glycosylation is a post-translational modification mediated by an enzymatic reaction catalysed by glycosyltransferases, which add a carbohydrate molecule to a predetermined region of a protein. More than 300 glycosyltransferases have been identified in mammals [1]. In contrast, glycation is a random nonenzymatic reaction that occurs under conditions of hyperglycaemia and ageing. The reactive reducing ends of free sugars (e.g., glucose, fructose, and galactose) covalently attach to the amino acid residue of the protein, thereby creating glycated products.

Glycation has been previously studied. Robert Lynn from the United Kingdom first reported that proteins and reducing sugars react during the beer-making process to form new compounds [2]. Subsequently, the French chemist Louis-Camille Maillard discovered that heating a mixed solution of amino acids and reducing sugars produced a brown compound [3]; this was the first report of the Maillard reaction or aminocarbonyl reaction, which is a nonenzymatic reaction between the amino group of an amino acid and carbonyl group of a reducing sugar (**Figure 1**).

In the early stages of the Maillard reaction, the imine produced by the nucleophilic reaction of the amino group and carboxyl group becomes a stable Amadori compound through Amadori rearrangement. The Amadori compound then undergoes a repeated polycondensation reaction with an amino compound using ozone or

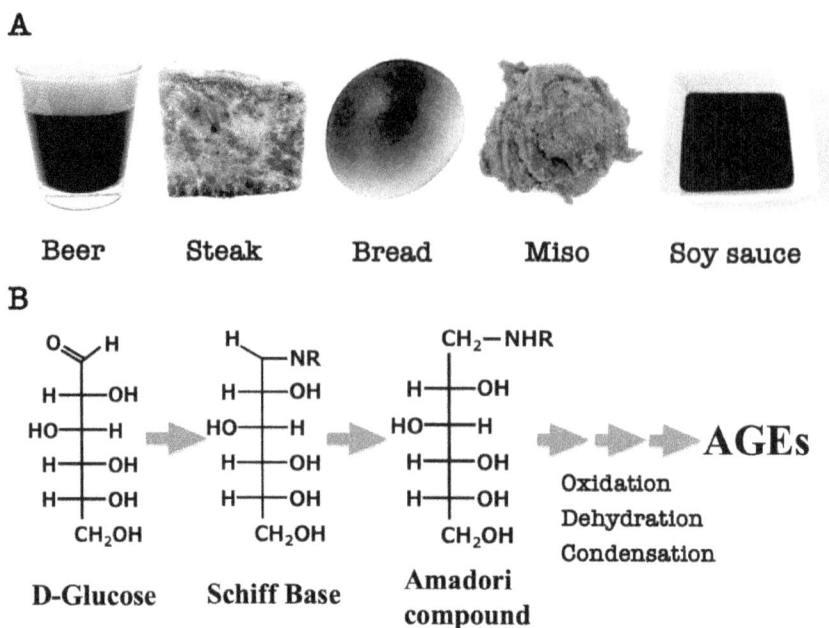

Figure 1.
Maillard reaction in foods and the formation of AGEs. (A) Proteins contained in foods are saccharified during fermentation and processing, and the Maillard reaction is accompanied by browning/denaturation. (B) The amino group of the amino acid of the protein and the carbonyl group of the reducing sugar react nonenzymatically, and AGEs are produced by repeating oxidation, dehydration, and condensation from the Schiff base via the Amadori compound.

furfural as an intermediate to produce a brown product, melainodin, in late stages [4]. Structures formed in the latter stage of the nonenzymatic glycation reaction between reducing sugars and proteins are collectively known as advanced glycation end products (AGEs).

Fermented foods, such as dark beer, miso, and soy sauce, contain large amounts of AGEs, including 3-deoxyglucosone and melanoidin [5]. Additionally, milk, cheese, and butter contain carboxymethyl lysine (CML) [6]. These chemicals are consumed on a daily basis and some AGEs, such as carbonyl compounds and CML, which are closely related to disease states, are known to be glycotoxins. Many studies have evaluated the adverse health effects of ingesting glycotoxins present in such foods in relation to nephropathy [7–9], type 2 diabetes [10, 11], and arteriosclerosis [12]; however, these relationships are not completely understood. Therefore, research on phytochemicals that prevent adverse effects on the living body caused by ingestion of these glycotoxins is being conducted [13–15].

In this chapter, we first introduce the biochemical properties of AGEs and their reaction processes. We then discuss intracellular signal transduction systems related to oxidative stress caused by AGEs in a hyperglycaemic environment and describe the relationships between AGEs and diseases.

2. Biochemical basis of AGEs

Protein glycation can be subdivided into three major stages: early, middle, and late. In the initial reaction, the carbonyl group (C=O) of a reducing sugar, such as glucose, reacts with the amino group (NH_2) of the amino acid residue in the protein to form a Schiff base (C=N). This Schiff base is relatively unstable and eventually becomes an enol, causing Amadori rearrangement and finally leading to the formation of a stable Amadori compound (C-N).

Kunkel found abnormal haemoglobin levels in the blood of normal people [16], and increased levels of abnormal haemoglobin were observed in patients with diabetes [17]. Currently, haemoglobin A1c (HbA1c), which is used as a diagnostic criterion for diabetes, is formed via Amadori rearrangement of the amino-terminal valine of the haemoglobin β chain and reflects the blood glucose level for 3–4 weeks [18, 19]. In the intermediate stage, α-dicarbonyl compounds, which are derivatives of sugars such as glucosone, 3-deoxyglucosone, glyoxal, and methylglyoxal, are produced from Amadori compounds. After further reacting with the amino compound, these α-dicarbonyl compounds undergo dehydration, condensation, cyclisation, and intermolecular crosslinking to form stable AGEs in the advanced stage (**Figure 2**). The pathway through which AGEs are produced from these series of Schiff bases via Amadori compounds and α-dicarbonyl compounds is known as the Hodge pathway [4]. In addition, the Namiki pathway, which produces glyoxal and glycolaldehyde, generates free radicals from Schiff bases without producing Amadori compounds [20].

Because the Schiff base is in a state in which it easily undergoes a secondary reaction with sugars and amino acids, dehydration, isomerisation, cleavage, cyclisation, and polymerisation can be repeated; the final products produced through these intermediates are extremely diverse. Therefore, the structures of many compounds are complicated, and most have not been identified. The structures of typical AGEs, such as CML, pyrarin, argpyrimidine, and pentosidine, have been reported (**Figure 2**).

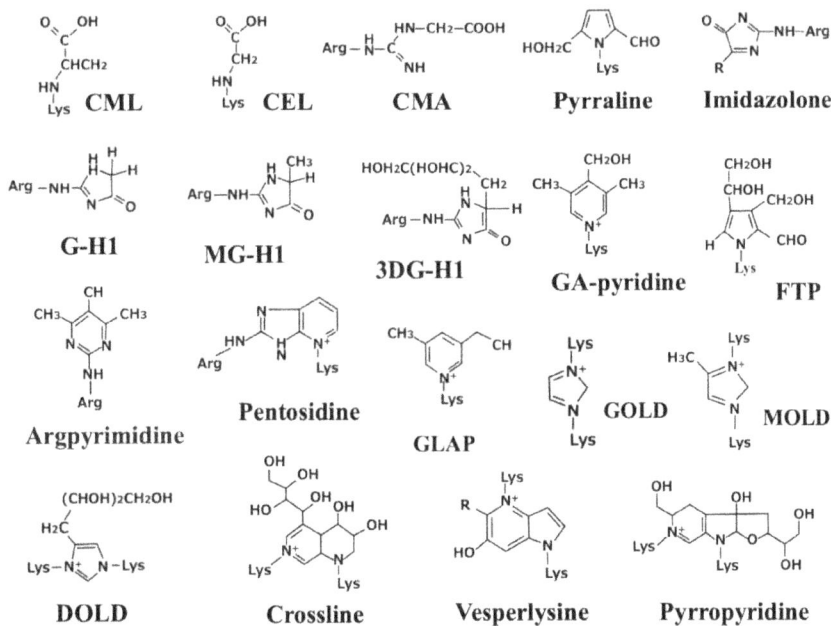

Figure 2.
The main chemical structures of AGEs. Abbreviations used: CML, Nε-carboxymethyl-lysine; CEL, Nε-(1-carboxyethyl)lysine; CML, Nω-(Carboxymethyl)-L-arginine; G-H1, Nδ-(5-hydro-4-imidazolon- 2-yl) ornithine; MG-H1, Nδ-(5-hydro-5-methyl-4-imidazolon-2-yl)-ornithine; 3DG-H1, Nδ-[5-(2,3,4-trihydroxybutyl)-5-hydro-4-imidazolon-2-yl] ornithine; GA- pyridine, Glycolaldehyde-pyridine; FTP, Formyl Threosyl Pyrrole; GLAP, glyceraldehyde-derived pyridinium-type advanced glycation end product; GOLD, glyoxal-derived lysine dimer; MOLD, methylglyoxal-derived lysine dimer; DOLD, 3-deoxyglucosone-derived lysine dimer

3. *In vivo* AGE generation pathways

To date, AGEs have been widely studied because of the close involvement in diabetic complications. HbA1c is currently used as a diagnostic criterion and indicator of mean blood glucose levels over a period of 1–2 months in patients with diabetes. Albumin, another representative protein in the blood, is also related to diabetic complications. In patients with diabetes, albumin has been shown to glycate four lysine residues (K199, K281, K439, and K525) in the molecule [21]. In addition, albumin is more easily saccharified than haemoglobin, and its reaction is rapid; thus, blood GA levels fluctuate more than HbA1c levels. Accordingly, gluco-albumin, which has a short half-life, was recently reported as an index of the average blood glucose level over a period of approximately 2 weeks [22].

At the experimental level, bovine serum albumin (BSA) has been used to evaluate the functions of AGEs *in vivo*. Various specific antibodies have been produced by immunisation with glycated AGE-BSA as antigens. Many commercially available AGEs are produced *in vitro* by incubating BSA and D-glucose at 37°C for 8 weeks in 0.2 M phosphate buffer (pH 7.4) and 5 mM DTPA. Farboud et al. reacted BSA with glycolaldehyde to produce pentosidine-BSA and obtained antibodies that recognise CML and pentosidine from this antigen [23]. Takeuchi named these six types of AGEs as glucose-derived AGE-1 (Glc-AGE), glyceraldehyde-derived AGE-2 (Glycer-AGE), glycol aldehyde-derived AGE-3 (Glycol-AGE), methylglyoxal-derived AGE-4 (MGO-AGE), glyoxal AGE-5 (GO-AGE), and 3-deoxyglucosone-derived AGE-6 (3DG-AGE); they then produced specific antibodies against each of the six types [24–26] (**Figure 3**). Using these antibodies, Takeuchi et al. clarified that AGE-2 derived from glyceraldehyde and AGE-3 derived from glycolaldehyde, produced by Schiff bases and Amadori compounds, were closely related to the onset and progression of diabetic retinopathy and nephropathy compared with AGE-1 [27–30]. The authors also demonstrated that these highly toxic AGE-2 and AGE-3 act via receptors for AGEs (RAGE) and therefore named these molecules toxic AGEs (TAGEs) [31], and identified nontoxic AGEs, including AGEs such as CML, pentocidin, and pyrrolin that are generated from glucose and by active trapping and detoxification of highly chemically reactive aldehyde/carbonyl compounds occurring in the body. TAGEs derived from glyceraldehyde, glycolaldehyde, and acetaldehyde are critical to the development and progression of various diseases and should be considered separately from other AGEs [32].

During the production of TAGEs, unique glucose metabolism pathways have been identified in the hyperglycaemic environment associated with diabetes. For example, in the hyperglycaemic environment observed in patients with type 2 diabetes, intracellular glucose levels are abnormally elevated in cells that take up insulin-independent glucose, such as the liver, brain, and placenta. The liver expresses the glucose transporter (GLUT) named as GLUT2, which has a low affinity for and takes up a large amount of glucose. GLUT3, which has a high affinity for glucose, also functions in glucose transport [33]. In such cells, the extra glucose is shunted into the polyol pathway by saturation of the normal glycolytic pathway [34, 35]. The polyol pathway is a side pathway that is activated when glycolysis is stagnant. First, excess glucose, which is not metabolised by glycolysis, is converted to sorbitol (polyol) by aldose reductase, after which sorbitol is metabolised to fructose by sorbitol dehydrogenase. When aldose reductase is enhanced, excessive consumption of its coenzyme NADPH causes a decrease in reduced glutathione and abnormalities in the active oxygen scavenging system. Such an increase in aldose

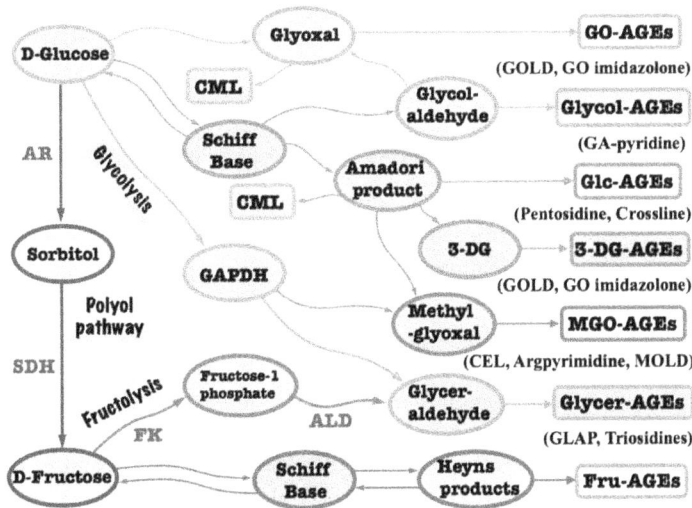

Figure 3.
AGE generation process in vivo. In the living body, AGEs are produced via dicarbonyl compounds generated during glucose metabolism of reducing sugars, such as glucose. In a hyperglycaemic environment, when glycolysis is stopped, the polyol circuit is enhanced, and glyceraldehyde-AGEs are produced. GO-AGEs, glyoxal (GO)-derived AGEs; glycol-AGEs, glycolaldehyde-derived AGEs; Glc-AGEs, glucose-derived AGEs; 3-DG-AGEs, 3-deoxyglucosone (3-DG)-derived AGEs; MGO-AGEs, methylglyoxal (MGO)-derived AGEs; glycer-AGEs, glyceraldehyde-derived AGEs; CML, Nε-(carboxymethyl) lysine. This figure has been modified based on the reference [25, 26].

reductase in type 2 diabetes is thought to worsen haemodynamics and lead to diabetic neuropathy (DN) [36]. Therefore, in patients with diabetes, the concentration of fructose produced from glucose is increased intracellularly because of enhancement of the polyol pathway [37, 38].

Fructose produced by this polyol pathway is thought to have a stronger protein glycation ability than glucose [39]. Therefore, increases in intracellular fructose promote AGE formation [40]. In our research, we attempted to suppress protein saccharification by inhibiting aldose reductase. Administration of the aldose reductase inhibitor Solvinyl to streptozotocin-induced diabetic rats reduced AGEs in skin collagen [41]. Moreover, the pentosidine-like fluorescence (335/385 nm) of the crystalline lens of galactosaemic rats was suppressed by treatment with the aldose reductase inhibitor sorbinin [42]. Administration of an aldose reductase inhibitor to patients with diabetes reduces the amount of N-epsilon-(carboxymethyl)-lysine in erythrocytes [43]. Following the development of many aldose reductase inhibitors, epalrestat was used clinically [44].

Fructose generated from such a polyol pathway is converted to fructose-1-phosphate by fructokinase, and fructose-1-phosphate further produces glyceraldehyde by aldolase. AGEs formed from this glyceraldehyde are highly toxic TAGEs. Increases in intracellular fructose, which trigger glyceraldehyde production, are caused not only by the polyol pathway but also by excessive intake of high-fructose syrup, such as high-fructose corn syrup.

Fructose is a natural ketose that is abundant in fruits and honey. However, in recent years, many soft drinks have been produced using high-fructose corn syrup, which is an isomerised sugar, and a relationship between excessive intake of fructose and metabolic syndrome has been reported [45]. Fructose ingested from soft drinks is taken up into cells by passive transport via GLUT5 in the epithelium of the small intestine. In contrast, glucose and lactose-derived galactose are taken up into cells by active transport via sodium-glucose cotransporter 1. Excessive fructose is transported from small intestinal epithelial cells through the portal vein to the liver and the whole body, thereby increasing glyceraldehyde-derived TAGEs. As discussed later, glyceraldehyde-derived TAGEs generated from fructose can cause liver diseases.

4. AGE receptors

Accumulation of AGEs *in vivo* causes a decrease in physiological function, leading to the onset and progression of various diseases. Recent studies revealed the existence of receptors involved in degrading and removing AGEs accumulated by glycation of such proteins and the intracellular signal transduction system via receptors [46]. AGEs are categorised into two groups based on their receptors; the first group includes the receptors AGE-R1, AGE-R3, scavenger receptor class A (SR-A) I, SR-AII, scavenger receptor-BI (SR-BI), cluster of differentiation 36 (CD36), FEEL1, FEEL2, and ezrin/radixin/moesin (ERM), which exert scavenger functions to removes AGE, and the second group includes RAGE, which is related to the enhancement of inflammation and oxidative stress (**Figure 4**).

AGE-R1 and AGE-R2 were identified as oligosaccharyltransferase-48 (OST-48) and 80-kDa protein kinase C (PKC) substrate (80 K-H), respectively, in rat livers [47]. Subsequently, AGE-R3 was identified as a protein that binds to AGE-1 and AGE-2 [48] to form a complex. AGE-R1 is also known as OST-48, belongs to the single transmembrane lectin family, and has a molecular weight of 48 kDa. AGE-R1 is expressed in endothelial cells, mesangial cells, macrophages, and mononuclear cells and functions by removing AGEs via endocytosis. AGE-R1, which enhances AGE removal, may also be a distinct receptor, as it suppresses AGE-mediated mesangial cell inflammatory injury by protecting against injury to the kidneys and other tissues due to diabetes [49]. Recent studies reported that AGE-R1 may be involved in lifespan extension [50, 51]. AGE-R2, also known as 80 K-H, is a tyrosine phosphorylated protein with a molecular weight of 80 kDa that was initially identified as a substrate for PKC and is expressed in the cytoplasm [47]. AGE-R2 is expressed

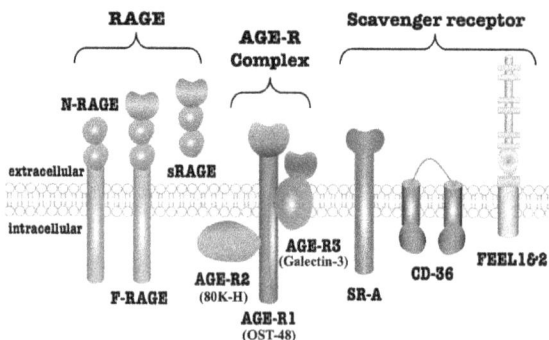

Figure 4.
The receptors for AGEs. A schematic diagram of AGE receptors is shown [46]. The receptor of AGEs (RAGE) includes full-length RAGE (F-RAGE), N-terminally truncated RAGE (N-RAGE), and soluble RAGE (sRAGE), which are cleaved from the cell surface membrane by matrix metalloproteinases. The AGE receptor (AGE-R complex) contains AGE-R1 (OST-48), AGE-R2 (80K-H), and AGE-R3 (Galectin-3). Scavenger receptor class A (SR-A), cluster of differentiation 36 (CD36), fasciclin EGF-like, laminin-type EGF-like, and link domain-containing scavenger receptor 1 and its homolog 2 (FEEL1 and − 2) are indicated as scavenger receptors.

in mononuclear cells and in the kidneys, vascular endothelium, brain, and nerves. Importantly, AGE-R2 is involved in activating intracellular signals via receptors, such as fibroblast growth factor receptor [52, 53]. AGE-R3, also called galectin-3, is a receptor that belongs to the lectin family and has a molecular weight of 32 kDa [48]. AGE-R3 binds directly to AGEs via the carbohydrate recognition domain in cells and is expressed in macrophages, eosinophils, and mast cells as well as in the nerves and kidneys. AGE-R3 has been reported to suppress adhesion between cells and the matrix laminin [54], activate mast cells [55], and degrade AGEs via endocytosis [48]. In addition, when diabetes develops in AGE-R3-knockout mice, the expression of macrophage scavenger receptor A and AGE-R1, which is involved in degrading AGEs, is decreased, and the expression of AGE receptors related to cell damage, such as RAGE and AGE-R2, is increased [56]. Because the expression of AGE-R3 is enhanced in ageing and diabetes, this receptor may have protective effects against ageing [57].

SR-A has been identified as a macrophage scavenger receptor [58, 59] and has a wide range of functions, such as removal of acetylated or oxidised low-density lipoprotein (LDL), removal of apoptotic cells, biological defence from bacteria, and cell adhesion [60]. SR-A is highly expressed in peritoneal macrophages derived from humans and from diabetic mice after culture in high-glucose medium [61]. Furthermore, SR-A promotes macrophage infiltration and foaming by incorporating AGEs into cells from the cell surface of macrophages [62, 63]. SR-BI is expressed in macrophages and in the liver adrenal glands and ovaries, functioning to promote the uptake of the cholesterol ester of high-density lipoprotein (HDL) and subsequent return of HDL to the liver [64, 65]. CD36, also known as scavenger receptor-BII, is a highly expressed receptor for single-stranded glycoprotein of 88 kDa in macrophages, vascular endothelial cells, and adipocytes [66]. CD36 binds to fatty acids, collagen, and oxidised LDL and is responsible for the uptake of oxidised LDL into macrophages and transport of fatty acids to adipocytes. Because CD36 is involved in removing AGEs, this protein may play protective roles in atherosclerotic diseases [67, 68]. The fasciclin, EFG-like, laminin-type EGF-like, and link domain-containing scavenger receptor-1 (FEEL-1) is expressed in the liver, vascular endothelial cells, and monocyte lineage cells, whereas FEEL-2 (a homologue of FEEL-1) is expressed in the spleen and lymph nodes. Despite the different tissue specificity, FEEL-1 and -2 are believed to be involved in the degradation of AGEs [69]. Megalin was identified as a 600-kDa glycoprotein (gp330) antigen expressed in glomerular epithelial cells (podocytes) of Heymann nephritis, a rat model of membranous nephropathy [70]. In recent studies, megalin was shown to bind to AGEs; AGEs that have passed through glomeruli are trapped and taken up by lysosomes to be decomposed [71]. AGEs bind to the N-terminus of the ERM protein family, which is a linker protein that crosslinks actin filaments and cell membrane proteins [72]. AGEs have been shown to promote angiogenesis through the hyperpermeability of human umbilical vein endothelial cells by inducing the phosphorylation of moesin via the RhoA/ROCK pathway [73].

RAGE is a single-pass 45-kDa transmembrane protein belonging to the immunoglobulin superfamily and was first isolated and identified from bovine lungs as a cell surface receptor that binds to AGEs [74]. RAGE is expressed in monocytes, macrophages, nerves, renal tubule cells, and mesangial cells [75]. In addition to AGEs, RAGE also binds to amyloid β protein, S100/calgranulins, and high-mobility group box 1 as ligands and is involved in the enhancement of inflammation and oxidative stress [76, 77]. RAGE is composed of a total of five domains: the extracellular domain of one V domain and two C domains, transmembrane domain, and intracellular domain [78]. When AGEs bind to this full-length RAGE, NADPH oxidase is activated, and the production of intracellular reactive oxygen species

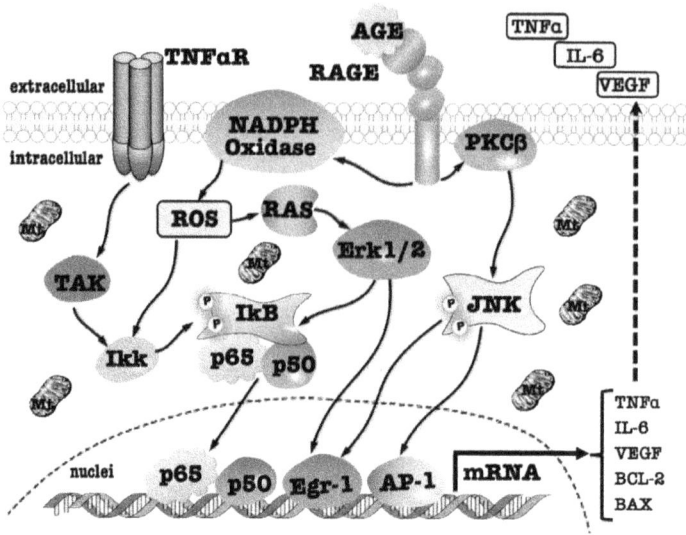

Figure 5.
AGE/RAGE signalling. NADPH oxidase is activated by the binding of AGE to RAGE, and intracellular ROS levels are elevated. Intracellular ROS activates the IκB kinase (IKK) complex and inhibitor of NF-κB (IκB), stimulating the translocation of the NF-κB subunits p65 and p50 and activating transcription. In addition, activation of PKCβ stimulates transcription via activator protein-1 (AP1) in the nucleus by phosphorylation of c-Jun N-terminal kinase (JNK). Enhancement of these inflammatory signals releases inflammatory cytokines, such as TNFα and IL-6, as well as VEGF, which is involved in angiogenesis, and B-cell lymphoma 2 (Bcl-2) and Bcl-2 associated X protein (Bax), which are involved in apoptosis. TNFα, an inflammatory cytokine, is released extracellularly and binds to the TNFα receptor, and activation of TGFβ activated kinase (TAK) reactivates JNK.

(ROS) is promoted [79, 80]. ROS upregulate various inflammatory cytokines, growth factors, and adhesion molecules by activating nuclear factor-kappa B (NF-κB) signalling. In addition, c-Jun N-terminal kinase (JNK), a major subfamily of ROS-activated mitogen-activated protein kinase pathways, has been shown to cause cell apoptosis and dysfunction (**Figure 5**) [81]. In addition to full-length RAGE on the cell surface, RAGE can be expressed as two splice variants, i.e., the intracellular domain-deficient type (C-terminally truncated RAGE) and extracellular V domain-deficient type (N-terminally truncated RAGE) [82]. Of these, the intracellular domain-deficient RAGE is called soluble RAGE (sRAGE). sRAGE can further be divided into endogenous secretory RAGE (esRAGE) and soluble RAGE, which are cleaved by proteases such as matrix metalloproteinases [83]. sRAGE has a binding site for AGEs and is thought to function as a decoy receptor that captures extracellular AGEs and inhibits binding to RAGE on the cell surface, thereby blocking intracellular signals [84]. Blood esRAGE levels are significantly lower in patients with type 2 diabetes than in patients without diabetes, suggesting that this target is involved in the development of type 2 diabetes [85]. Moreover, blood esRAGE levels in patients with type 2 diabetes are inversely correlated with the severity of carotid atherosclerosis and coronary artery disease as complications [86, 87].

5. AGEs and oxidative stress

Intracellular signal transduction of AGEs via RAGE increases intracellular ROS. ROS are oxygen-containing molecular derivatives that are in a more activated state than triplet oxygen, which is a ground-state oxygen molecule necessary for normal biological activities and is highly reactive, resulting in oxidative damage to various biological components. The main active oxygen species are singlet oxygen, superoxide, hydrogen peroxide, and hydroxyl radicals [88]; these molecules react with biopolymers, such as DNA, lipids, proteins, and enzymes, resulting in lipid peroxidation, DNA mutations, protein denaturation, and enzyme inactivation. Many amino acids are carbonylated and modified by ROS for detection of protein carbonylation using mass spectrometers [89]. Moreover, carbonylation of this protein is caused by addition reaction of aldehydes because of the peroxidation reaction of lipids and saccharification reaction of proteins described above [90, 91]. Highly reactive α-dicarbonyl compounds, such as 3-deoxyglucosone (3-DG), glyceraldehyde, and methylglyoxal, are produced from the Amadori compound generated by saccharification [91]. These AGEs then recombine with RAGE, creating a vicious cycle in which more ROS are generated. Such ROS are considered to have negative effects because overproduction of ROS is closely associated with ageing due to oxidative stress, cancer, and the development of lifestyle-related diseases [91]. However, ROS (e.g., superoxide and hydrogen peroxide) produced by white blood cells play important roles in biological defence and immune function [92]. ROS are also used in a wide range of tissues and cells as bioactive substances for intracellular signal transduction, fertilisation, cell differentiation, and apoptosis [93].

Because glucose is metabolised to obtain energy, the carboxyl group of glucose reacts with the amino group of the protein during the metabolic process to form AGEs in the body nonenzymatically via the Amadori compound. With ageing, these AGEs accumulate in various organs in the body, resulting in oxidative stress, ROS generation, and progression of organ stress. Thus, ageing is related to oxidative stress induced by AGEs. Additionally, AGEs-ised HbA1c levels in the blood have been used as an index for controlling blood glucose levels in clinical practice for patients with diabetes. Kusunoki et al. showed that fasting serum 3-DG levels in patients with diabetes were significantly higher than those in controls. Additionally, serum 3-DG levels tended to be higher in patients with diabetes showing low nerve conduction velocity [94]. In patients with diabetes, AGEs generated from excess glucose circulate throughout the body via the blood and increase oxidative stress in various organs. Therefore, in the hyperglycaemic environment associated with diabetes, oxidative stress due to excess glucose is thought to be significantly involved in the development of diabetic complications.

6. AGEs and diabetic complications

Hyperglycaemia in diabetes mellitus affects many organ systems, including the eyes, kidneys, heart, and peripheral and autonomic nervous systems. They can be broadly divided into microangiopathy, which occurs mainly in the capillaries, and macroangiopathy, which occurs in relatively large blood vessels. Three major complications, i.e., diabetic retinopathy, diabetic nephropathy, and DN, are microangiopathies that occur in patients with diabetes [95]. In contrast, arteriosclerotic diseases, which cause vascular diseases, such as myocardial infarction and cerebral infarction, are considered as macroangiopathies. AGEs are the leading causes of complications caused by microangiopathy and macroangiopathy [96–98].

Diabetic retinopathy causes bleeding and ischaemia in capillaries due to the hyperglycaemic environment, and progression results in bleeding or retinal detachment inside the vitreous body. AGEs are associated with the presence and progression of diabetic retinopathy [99]. Diabetic keratopathy, in which the corneal epithelium is exfoliated due to aggregation of AGEs-ised proteins, is thought to be related to AGE formation via laminin, which is found in the basement membrane of the corneal epithelium [100]. In human RAGE transgenic mice induced by streptozotocin as an experimental model of diabetes, the blood-retinal barrier was disrupted, and leukostasis was increased [101]. However, systemic administration of sRAGE intraperitoneally suppressed collapse of the blood-retinal barrier and leukostasis [101]. Administration of soluble RAGE, which comprises the extracellular domain of RAGE, enhances AGEs in the blood and blocks the interaction with cell membrane RAGE. As a result, pathological conditions related to diabetic retinopathy, such as increased retinal vascular permeability and adhesion of leukocytes to retinal blood vessels, can be suppressed [101, 102]. Thus, AGE/RAGE signalling plays important roles in the development of diabetic retinopathy.

The kidney is an organ that filters waste products in the blood to produce urine and is formed by the renal glomerulus, which is similar to a mass of capillaries. In patients with diabetes, renal dysfunction can also occur. Chronic kidney disease occurs in approximately 20–40% of patients with diabetes [103]. If renal failure occurs, artificial haemodialysis is required. Diabetic nephropathy is the most common cause of dialysis. In diabetic nephropathy, accumulation of AGEs has been reported in various cells, such as the glomerular basement membrane, mesangium, podocytes, tubular cells, and endothelial cells [104]. In addition, several studies have suggested that RAGE expression is increased in patients with diabetic nephropathy [104, 105]. Administration of AGEs to nondiabetic rats induces proteinuria and degenerative changes in the renal tissue, highlighting the important roles of AGEs in the development of diabetic nephropathy [106]. CML in patients with type 1 diabetes was found to correlate with the severity of nephropathy [107]. Moreover, the levels of CML- and hydroimidazolone-AGEs in the serum of patients with type 2 diabetes are significantly increased [108]. CML-human serum protein levels are higher in patients with proteinuria, and increased levels of circulating AGE peptides are correlated with the severity of renal dysfunction [109]. Studies in RAGE transgenic mice revealed the development of advanced diabetic nephropathy features, such as renal hypertrophy, glomerular hypertrophy, mesangial enlargement, glomerulosclerosis, and proteinuria [110]. In OVE26 mice, a diabetic mouse model that exhibits progressive glomerular sclerosis and decreased renal function, RAGE deficiency alleviates histological and morphological changes and albuminuria associated with diabetic nephropathy and does not result in decreased renal function [111]. Thus, these findings support that RAGE is involved in the development of diabetic nephropathy and as a target molecule in for treating this disease.

DN is a peripheral nerve disorder caused by prolonged hyperglycaemia in diabetes, resulting in numbness, pain, and hypoesthesia of the limbs. In the nervous tissue, hyperglycaemia increases non-insulin-dependent glucose uptake. Excess glucose is thought to cause sorbitol accumulation via the polyol pathway and microangiopathy, which nourishes the nerves. Accumulation of AGEs is observed in perineurial cells, nerve axons, and Schwann cells in the peripheral nerves of patients with diabetes [112]. In Schwann cells, neurofilaments and tubulin, which are important for axonal transport, are converted to AGEs [113]. Overexpression of AGEs and RAGE in the nerves of patients with diabetes activates NF-κB; these changes correlate with hypoesthesia [114]. Therefore, antiglycation agents, such as aminoguanidine, have been promoted as treatments for DN [115]. However, aminoguanidine was shown to have various side effects in a clinical trial of patients with DN, and thus its development was discontinued. Recently, the anti-inflammatory cytokine interleukin-10 has attracted attention because of ability to suppress AGE-induced apoptosis in Schwann cells by reducing oxidative stress through inhibition of NF-κB activation [116]. Thus, the potential use of interleukin-10 for treating DN is also being discussed.

7. AGEs and arteriosclerosis

In addition to the three major complications of diabetes (i.e., diabetic retinopathy, diabetic nephropathy, and DN), if hyperglycaemia continues for a long time, ischaemic heart disease, cerebral infarction, and macroangiopathy (peripheral arterial disease progression) can occur due to arteriosclerosis in large blood vessels, such as the heart and brain. Inflammation in the blood vessel wall is critical for the onset and progression of arteriosclerosis. AGEs produced in a hyperglycaemic environment bind to RAGE in vascular endothelial cells and activate AGE/RAGE signalling. As a result, the expression of inflammatory cytokine genes is enhanced by NF-κB signalling and the phosphorylation of JNK because of the production of ROS by NADPH oxidase, causing inflammation of the blood vessel wall [117]. Recent studies showed that vascular endothelial growth factor is involved in increases in atheroma in atherosclerotic lesions [118]. Moreover, AGEs induce angiogenesis by promoting the production of vascular endothelial growth factor autocrine signalling in endothelial cells, enhancing inflammation in blood vessels, and increasing atheroma [117]. Excess sRAGE has been reported to inhibit AGE/RAGE signalling and suppress the onset and progression of arteriosclerosis [119–121]. Furthermore, AGEs have been detected in cultures of mouse or human aortic endothelial cells in a hypoxic state, suggesting that RAGE signalling is activated by hypoxia in aortic endothelial cells [122]. Early growth response-1 expression under hypoxic conditions, PKC translocation, and JNK phosphorylation are inhibited by sRAGE or anti-AGE antibodies, and *RAGE* is downregulated by aminoguanidine and siRNA.

8. AGEs and intrauterine hyperglycaemia

In pregnant women or those with gestational diabetes during pregnancy, hyperglycaemia can create a hyperglycaemic environment in the uterus through the placenta. However, few studies have evaluated the molecular mechanisms by which the intrauterine hyperglycaemic environment affects foetal development and future illnesses in offspring. One study evaluated the hearts of infants born from diabetic pregnancy model rats with hyperglycaemia during pregnancy [123]. Additionally, a gestational diabetes rat model was created by administration of streptozotocin via the tail vein immediately after pregnancy. Akt-related insulin signalling was abnormal in the hearts of offspring born to mothers of these gestational diabetes model rats [124]. We investigated the expression of the insulin signalling system, ROS, AGEs, and related genes in the hearts of infants and in primary myocardial cultured cells (cardiomyocytes) isolated from the heart [125]. In primary cardiomyocytes isolated from the hearts of infants born to mothers with diabetes, insulin stimulation inhibited the translocation of GLUT4 to the cell membrane, indicating that insulin resistance was induced. Moreover, various proteins were excessively AGE-ised in the hearts and cardiomyocytes of offspring born from diabetic mother rats [125]. Intracellular ROS levels and *NF-κB*, tumour necrosis factor (*TNFα*), and *IL-6* gene expression levels in isolated cardiomyocytes were significantly increased compared with those in offspring of normal mother rats [125]. Thus, in offspring who spent the foetal period in an intrauterine hyperglycaemic environment, maternal hyperglycaemia may have caused abnormal insulin signalling due to the chronic inflammation induced by intracellular ROS and excessive AGE formation, thereby leading to cardiac hypertrophy [125]. Interestingly, daily oral administration of the n-3 unsaturated fatty acid eicosapentaenoic acid by gastric sonde to mother rats ameliorated this abnormal signal transduction in the

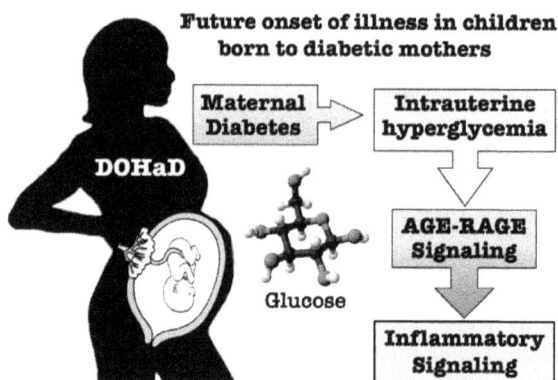

Figure 6.
The risk of future illness in children born to diabetic mothers. In diabetic mothers, maternal hyperglycaemia creates a hyperglycaemic environment in the womb through the placenta. During this time, the foetus is exposed to hyperglycaemia, and excessive hyperglycaemia activates AGE/RAGE signalling. This can cause the foetus to be exposed to an inflammatory cytokine storm. In addition, many proteins and enzymes are denatured by oxidative stress, which can also affect foetal development, and these effects may lead to the onset of disease after birth. Therefore, glycaemic control during pregnancy is critical.

heart. Based on these findings, the intrauterine hyperglycaemic environment of pregnant women may have major effects on various organs other than the heart in children through oxidative stress caused by excessive AGEs, including AGE/RAGE signalling. In addition, the intrauterine hyperglycaemic environment may affect offspring through epigenetics [125, 126].

The concept that malnutrition in the womb may affect the future development of lifestyle-related diseases in children was first proposed by David Barker of Southampton University in the 1980s [127]. Barker and colleagues used birth weight as an indicator of foetal nutrition and examined its association with various causes of death; their results showed that children born with a low birth weight were at high risk of dying from heart disease in the future [128]. Birth cohort studies have reported a series of epidemiological studies supporting the theory of adult disease foetal onset, including the fact that foetuses exposed to malnutrition may develop lifestyle-related diseases in adulthood [129] by inducing an adaptive response that predicts the future environment by regulating gene expression [130]. Peter Gluckman, Mark Hanson, and others further developed this theory of adult disease foetal onset into a generalised theory on the developmental origins of health and disease [131]. However, in modern society, eating habits have changed dramatically, and overnutrition, including obesity and diabetes, has become a challenge. Importantly, oxidative stress caused by exposure to the maternal hyperglycaemic environment may also have major effects on the future onset of illness in offspring (**Figure 6**).

9. Development of therapeutic agents targeting the AGEs-RAGE system

As described above, in a hyperglycaemic environment, oxidative stress induced by AGEs and RAGE can induce the onset and progression of various diabetic complications; hence targeting the AGEs-RAGE system, using AGEs formation inhibitors, AGEs degrading agents, AGEs-RAGE inhibitors and signal transduction inhibitors, may be an effective treatment strategy.

The first reported AGEs formation inhibitors are aminoguanidine and OPB-9195 (2-isopropylidenehydrazono-4-oxo-thiazolidine-5-ylacetanilide) which can capture reactive carbonyl compounds such as methylglyoxal and 3-DG and inactivate metal ions that catalyse radical formation such as chelating agents [132–134]. OPB-9195 has a stronger AGEs formation inhibitory activity than aminoguanidine [135], however, these compounds are associated with side effects such as vitamin B6 deficiency due to the capture of pyridoxal phosphate, anaemia, and liver damage, therefore, their clinical application has been discontinued. LR-90 (methylene bis [4,4-(2 chlorophenylureido phenoxyisobutyric acid)]) and ALT946 (N-(2-acetamidoethyl) hydrozinecarboximidamide hydrochlolide) are more potent AGEs inhibitors than aminoguanidine and OPB-9195 [136, 137], and are associated with fewer side effects; in particular, ALT946 has no NO synthase inhibitory activity, which is a side effect of aminoguanidine [137].

Pyridoxamine, a vitamin B6, has been reported to have renal damage-suppressing effects as well as carbonyl compound capturing and antioxidant effects [138–140]. Benfophothiamine, a vitamin B1 derivative, has various effects such as inhibiting AGEs formation, suppressing PKC activity and oxidative stress, activating transketolase, and inhibiting the polyol pathway [141]. Furthermore, sorbinin inhibits AGEs formation by blocking the polyol pathway [41, 42]. The renal protective effect of the renin-angiotensin system targeting drugs is attributed to the inhibition of pentosidine production [142]. The oral hypoglycaemic agent metformin inhibits AGEs formation via carbonyl compound capturing, metal chelate formation, and antioxidant activity [143].

N-phenacylthiazolium bromide (PTB) can cleave protein cross-linked by AGEs [144]. PTB water solubility increases when it is in the form of 3-phenacyl-4,5-dimethylthiazorium chloride (ALT-711). ALT-711 has been reported to suppress the accumulation of AGEs and improve vascular hardening and systolic blood pressure [145]. PTB and ALT-711 are therefore referred to as AGEs breaker agents. Certain plant extracts have been reported to exhibit this anti-AGEs effect. For example, terpinen-4-ol of citron (*Citrus junos*) has also been reported to decompose AGEs [146]. In addition, RAGE antagonists that block the interaction between AGEs and RAGE have been extensively studied [147].

Drugs targeting the AGEs-RAGE system primarily include AGEs formation inhibitors, AGEs breakers, and AGEs-RAGE signal inhibitors, which are investigated in non-clinical studies. Presently, the agents used for targeting AGEs-RAGE system in clinical settings include aldose reductase inhibitors, renin-angiotensin-based active drugs, and metformin. The reason behind using such diverse drugs and difficulty in discovering a specific drug is attributed to the structural diversity of AGEs, the multi-ligand receptor characteristics of RAGE, and the limited underdamping of the condition in which oxidative stress is generated in cells. However, oxidative stress induced by AGEs in a hyperglycaemic environment significantly influences the onset and progression of several lifestyle-related diseases. Therefore, advance translational research is essential to tackle challenges that basic research cannot.

10. Conclusions

As discussed in this chapter, glycation is a random, nonenzymatic reaction that differs significantly from enzymatically catalysed glycosylation. AGEs formed by saccharification consist of a wide variety of molecular species, many of which have not been structurally characterised, and these species vary from harmful to harmless. Oxidative stress, including ROS, is induced by AGEs during normal metabolism but is mitigated physiologically by antioxidant enzymes in the body. However, in a hyperglycaemic environment, as is typically observed in patients with diabetes, oxidative stress that cannot be removed via the antioxidant system of the body causes various diabetic complications such as organ stress. As the population of patients with diabetes continues to increase, the number of pregnant women with diabetes is also increasing due to late marriage and an older age of primigravida. Research results have strongly supported that the maternal hyperglycaemic state creates an intrauterine hyperglycaemic environment through the placenta that is involved in the development of various diseases in the offspring. Further studies are needed to clarify the molecular mechanism involved in oxidative stress and disease caused by glycation and to link these mechanisms with the diagnosis and prevention of lifestyle-related diseases.

Acknowledgements

We gratefully acknowledge the work of past and present members of our laboratory. This work was supported in part by JSPS KAKENHI Grants (nos. 20 K11611, 15 K00809, and 18 K11136 to AN and RK), Dairy Products Health Science Council and Japan Dairy Association (to AN), and Research Program of Jissen Women's University (to AN).

Conflict of interest

The authors declare no conflicts of interest.

Author details

Akio Nakamura[1*] and Ritsuko Kawaharada[2]

1 Department of Molecular Nutrition, Faculty of Human Life Sciences, Jissen Womens University, Hino, Tokyo, Japan

2 Department of Health and Nutrition, Takasaki University of Health and Welfare, Takasaki, Gunma, Japan

*Address all correspondence to: nakamura-akio@jissen.ac.jp

IntechOpen

References

[1] Taniguchi N, Honke K, Fukuda M, et al. (eds.): Handbook of glycosyltransferases and related genes, 2 ed. Springer Japan 2014.

[2] Ling AR. Malting. Journal of the Institute of Brewing. 1908;14:494-521

[3] Maillard LC. Action des acides aminés sur les sucres; formation des méla-noidines par voie methodique. Comptes Rendus de l'Académie des Sciences, 1912;154:66-68.

[4] Hodge J E. Dehydrated foods: chemistry of browning reactions in model systems. Journal of Agricultural and Food Chemistry. 1953;1:928-43.

[5] Nomi Y, Annaka H, Sato S, et al. Simultaneous Quantitation of Advanced Glycation End Products in Soy Sauce and Beer by Liquid Chromatography-Tandem Mass Spectrometry without Ion-Pair Reagents and Derivatization. Journal of Agricultural and Food Chemistry. 2016;64:8397-8405. doi: 10.1021/acs.jafc.6b02500.

[6] Assar SH, Moloney AC, Lima M, et al. Determination of N ε-(carboxymethyl)lysine in food systems by ultra-performance liquid chromatography-mass spectrometry. Amino Acid. 2009;36:317-326.

[7] Zheng F, He C, Cai W, et al. Prevention of diabetic nephropathy in mice by a diet low in glycoxidation products. Diabetes Metabolism Research and Reviews. 2002; 18:224-237.

[8] Vlassara H, Cai W, Crandall J, et al. Inflammatory mediators are induced by dietary glycotoxins, a major risk factor for diabetic angiopathy. Proceedings of the National Academy of Sciences of the United States of America. 2002;9:15596-15601.

[9] Uribarri J, Melpomeni P, Cai W, Dietary glycotoxins correlate with circulating advanced glycation end product levels in renal failure patients. American Journal of Kidney Diseases. 2003;42:532-538.

[10] Melpomeni P, Teresia G, Weijing C, et al. Glycotoxins: A Missing Link in the "Relationship of Dietary Fat and Meat Intake in Relation to Risk of Type 2 Diabetes in Men" Diabetes Care. 2002;25:1898-1899.

[11] Uribarri J, Cai W, Ramdas M, et al. Restriction of advanced glycation end products improves insulin resistance in human type 2 diabetes: potential role of AGER1 and SIRT1. Diabetes Care. 2011;34:1610-1616.

[12] Lin RY, Choudhury RP, Cai W, et al. Dietary glycotoxins promote diabetic atherosclerosis in apolipoprotein E-deficient mice. Atherosclerosis. 2003;168 : 213-220.

[13] Wu C-H, Huang SM, Lin J-A, et al. Inhibition of advanced glycation end product formation by foodstuffs. Food and Function. 2011;2:224-234.

[14] Peng X, Ma J, Chen F, et al. Naturally occurring inhibitors against the formation of advanced glycation end-products. Food Function. 2011;2:289-301.

[15] Anwar S, Khan S, Almatroudi A, et al. A review on mechanism of inhibition of advanced glycation end products formation by plant derived polyphenolic compounds. Molecular Biology Reports. 2021 Jan 3. doi: 10.1007/s11033-020-06084-0. Online ahead of print.

[16] Kunkel HG, WAllenius G, New hemoglobin in normal adult blood. Science. 1955;122:288. doi: 10.1126/science.122.3163.288.

[17] Rahbar S. An abnormal hemoglobin in red cells of diabetics. Clinica Chimica

Acta. 1968;22:296-298. doi: 10.1016/0009-9981(68)90372-0.

[18] Koenig RJ, Blobstein SH, Cerami A. Structure of carbohydrate of hemoglobin AIc. Journal of Biological Chemistry. 1977;252:2992-2997.

[19] Koenig RJ, Peterson CM, Jones RL, et al. Correlation of glucose regulation and hemoglobin AIc in diabetes mellitus. The New England Journal of Medicine. 1976;295:417-420. doi: 10.1056/NEJM197608192950804.

[20] Namiki M, Hayashi T. Role of sugar fragmentation in an early stage browning of amino-carbonyl reaction of sugar with amino acid. Agricultural and biological chemistry. 1986;50:1965-1970.

[21] Day JF, Ingebretsen CG, Ingebretsen WR Jr, et al. Nonenzymatic glycosylation of serum proteins and hemoglobin: response to changes in blood glucose levels in diabetic rats. Diabetes. 1980;29:524-527. doi: 10.2337/ diab.29.7.524.

[22] Kennedy AL, Merimee TJ. Glycosylated serum protein and hemoglobin A1 levels to measure control of glycae- mia. Annals of Internal Medicine. 1981;95:56-58.

[23] Farboud B, Aotaki-Keen A, Miyata T, et al. Development of a polyclonal antibody with broad epitope specificity for advanced glycation end products and localization of these epitopes in Bruch's membrane of the aging eye. Molecular Vision. 1999;5:11.

[24] Takeuchi M, Makita Z, Bucala R, et al. Immunological evidence that non-carboxymethyllysine advanced glycation end-products are produced from short chain sugars and dicarbonyl compounds in vivo. Molecular Medicine. 2000;6:114-125.

[25] Takeuchi M, Yanase Y, Matsuura N, et al. Immunological detection of a novel advanced glycation end-product. Molecular Medicine. 2001;7:783-791

[26] Takeuchi M. Toxic AGEs (TAGE) theory: a new concept for preventing the development of diseases related to lifestyle. Diabetology & Metabolic Syndrome. 2020;30;12:105. doi: 10.1186/ s13098-020-00614-3.

[27] Yamagishi S, Amano S, Inagaki Y, et al. Advanced glycation end products-induced apoptosis and overexpression of vascular endothelial growth factor in bovine retinal pericytes. Biochemical and Biophysical Research Communications. 2002;290:973-978. doi: 10.1006/bbrc.2001.6312.

[28] Okamoto T, Yamagishi S, Inagaki Y, et al. Angiogenesis induced by advanced glycation end products and its prevention by cerivastatin. FASEB Journal. 2002;16:1928-1930. doi: 10.1096/fj.02-0030fje.

[29] Yamagishi S, Inagaki Y, Okamoto T, et al. Advanced glycation end product-induced apoptosis and overexpression of vascular endothelial growth factor and monocyte chemoattractant protein-1 in human-cultured mesangial cells. Journal of Biological Chemistry. 2002;277:20309-20315. doi: 10.1074/jbc. M202634200.

[30] Yamagishi S, Inagaki Y, Okamoto T, et al. Advanced glycation end products inhibit de novo protein synthesis and induce TGF-beta overexpression in proximal tubular cells. Kidney International. 2003;63:464-473. doi: 10.1046/j.1523-1755.2003.00752.x.

[31] Yonekura H, Yamamoto Y, Sakurai S, et al. RAGE engagement and vascular cell derangement by short chain sugar-derived advanced glycation end products. In: The Maillard reaction in food chemistry and medical science: update for post-genomic era (Excerpta Medica International Congress Series 1245), Horiuchi S, Taniguchi N,

Hayase F, Kurata T, Osawa T, editors. Amsterdam, The Netherland: Elsevier Science B.V. 2002 p129-135.

[32] Takeuchi M, Yamagishi S. TAGE (toxic AGEs) hypothesis in various chronic diseases. Medical Hypotheses 2004;63:449-452. doi: 10.1016/j.mehy.2004.02.042.

[33] Thorens B, and Mueckler M. Glucose transporters in the 21st Century. American Journal of Physiology-Endocrinology and Metabolism. 2010;298:141-145

[34] Gabbay KH. Hyperglycemia, polyol metabolism, and complications of diabetes mellitus. Annual Review of Medicine. 1975;26:521-536.

[35] Cheng HM, Gonzalez RG. The effect of high glucose and oxidative stress on lens metabolism, aldose reductase, and senile cataractogenesis. Metabolism. 1986;35:10-14, 1986.

[36] Pang L, Lian X, Liu H, et al. Understanding Diabetic Neuropathy: Focus on Oxidative Stres. Oxidative Medicine and Cellular Longevity. 2020, Article ID 9524635;13.

[37] Travis SF, Morrison AD, Clements RS Jr, et al. Metabolic alterations in the human erythrocyte produced by increases in glucose concentration. The role of the polyol pathway. Journal of Clinical Investigation. 1971;50:2104-2112. doi: 10.1172/JCI106704.

[38] Dyck PJ, Zimmerman BR, Vilen TH, et al. Nerve glucose, fructose, sorbitol, myo-inositol, and fiber degeneration and regeneration in diabetic neuropathy. New England Journal of Medicine. 1988;319:542-548. doi: 10.1056/NEJM198809013190904.

[39] McPherson JD, Shilton BH, Walton DJ. Role of fructose in glycation and cross-linking of proteins. Biochemistry. 1988;27:1901-1907. doi: 10.1021/ bi00406a016.

[40] Takagi Y, Kashiwagi A Tanaka Y, et. al. Significance of fructose-induced protein oxidation and formation of advanced glycation end product. Journal of Diabetes and its Complications. 1995;9:87-91

[41] Suárez G, Rajaram R, Bhuyan KC, et al. Administration of an aldose reductase inhibitor induces a decrease of collagen fluorescence in diabetic rats. Journal of Clinical Investigation. 1988; 82: 624-627.

[42] Nagaraj RH, Prabhakaram M, Ortwerth BJ, et al. Suppression of Pentosidine Formation in Galactosemic Rat Lens by an Inhibitor of Aldose Reductase. Diabetes. 1994; 43:580-586. doi: 10.2337/diab.43.4.580.

[43] Hamada Y, Nakamura J, Naruse K, et al. Epalrestat, an aldose reductase ihibitor, reduces the levels of Nepsilon-(carboxymethyl)lysine protein adducts and their precursors in erythrocytes from diabetic patients. Diabetes Care. 2000; 23:1539-1544.

[44] Ohmura C, Watada H, Azuma K, et al. Aldose Reductase Inhibitor, Epalrestat, Reduces Lipid Hydroperoxides in Type 2 Diabetes. Endocrine Journal 2009;56:149-156.

[45] Hannou SA, Haslam DE. Fructose metabolism and metabolic disease. Journal of Clinical Investigation. 2018;128:545-555. doi: 10.1172/JCI96702

[46] Ott C, Jacobs K, Haucke E, et al. Role of advanced glycation end products in cellular signaling. Redox Biology. 2014;2:411-429

[47] Li YM, Mitsuhashi T, Wojciechowicz D, et al. Molecular identity and cellular distribution of advanced glycation end product receptors: relationship of p60 to OST-48 and p90 to 80K-H membrane proteins. Proceedings of the National Academy of Sciences of the United States of

America. 1996;93:11047-11052. doi: 10.1073/pnas.93.20.11047.

[48] Vlassara H, Li YM, Imani, F, et al. Identification of Galectin-3 As a High-Affinity Binding Protein for Advanced Glycation End Products (AGE): A New Member of the AGE-Receptor Complex. Molecular Medicine 1995;1:634-646

[49] Lu C, He JC, Cai W, et al. Advanced glycation endproduct (AGE) receptor 1 is a negative regulator of the inflammatory response to AGE in mesangial cells. Proceedings of the National Academy of Sciences of the United States of America. 2004;101:11767-11772. doi: 10.1073/pnas.0401588101.

[50] Zhuang A, Forbes JM. Diabetic kidney disease: a role for advanced glycation end-product receptor 1 (AGE-R1)? Glycoconjugate journal. 2016;33:645-652. doi: 10.1007/s10719-016-9693-z.

[51] Cai W, He JC, Zhu L, et al. Coronary Heart Disease High Levels of Dietary Advanced Glycation End Products Transform Low-Dersity Lipoprotein Into a Potent Redox-Sensitive Mitogen-Activated Protein Kinase Stimulant in Diabetic Patients. Circulation. 2004; 110: 285-291.

[52] Goh KC, Lim YP, Ong SH, et al. Identification of p90, a prominent tyrosine-phosphorylated protein in fibroblast growth factor-stimulated cells, as 80K-H. Journal of Biological Chemistry. 1996;271:5832-5838. doi: 10.1074/jbc.271.10.5832.

[53] Forough R, Lindner L, et al. Elevated 80K-H protein in breast cancer: a role for FGF-1 stimulation of 80K-H. The International Journal of Biological Markers. 2003;18:89-98. doi: 10.5301/jbm.2008.563.

[54] Ochieng J, Leite-Browning ML, Warfield P, et al. Regulation of cellular adhesion to extracellular matrix proteins by galectin-3. Biochemical and Biophysical Research Communications. 1998;246:788-791. doi: 10.1006/bbrc.1998.8708.

[55] Hsu DK, Zuberi RI, Liu F-T, et al. Biochemical and biophysical characterization of human recombinant IgE-binding protein, an S-type animal lectin. Liu FT. Journal of Biological Chemistry. 1992;267:14167-1474.

[56] Pugliese G, Pricci F, Leto G, et al. The Diabetic Milieu Modulates the Advanced Glycation End Product–Receptor Complex in the Mesangium by Inducing or Upregulating Galectin-3 Expression: Diabetes. 2000; 49: 1249-1257

[57] Pugliese G, Pricci F, Iacobini C, et al. Accelerated diabetic glomerulopathy in galectin-3/AGE receptor 3 knockout mice. FASEB Journal. 2001;15:2471-2479. doi: 10.1096/fj.01-0006com.

[58] Matsumoto A, Naito M, Itakura H, et al. Human macrophage scavenger receptors: Primary structure, expression, and localization in atherosclerotic lesions. Proceedings of the National Academy of Sciences of the United States of America. 1991;87:9133-9137.

[59] Kodama T, Freeman M, Rohrer L, et al. Type I macrophage scavenger receptor contains alpha-helical and collagen-like coiled coils. Nature. 1990;343:531-535.

[60] Platt N, Gordon S. Is the class A macrophage scavenger receptor (SR-A) multifunctional? - The mouse's tale. Journal of Clinical Investigation. 2001;108:649-654. doi: 10.1172/JCI13903.

[61] Fukuhara-Takaki K, Sakai M, Sakamoto Y, et al. Expression of class A scavenger receptor is enhanced by high glucose in vitro and under diabetic

conditions in vivo: one mechanism for an increased rate of atherosclerosis in diabetes. Journal of Biological Chemistry. 2005;280:3355-3364. doi: 10.1074/jbc.M408715200.

[62] Suzuki H, Kurihara Y, Takeya M, et al. A role for macrophage scavenger receptors in atherosclerosis and susceptibility to infection. Nature. 1997;386:292-296. doi: 10.1038/386292a0

[63] Nagai R, Matsumoto K, Ling X, et al. Glycolaldehyde, a reactive intermediate for advanced glycation end products, plays an important role in the generation of an active ligand for the macrophage scavenger receptor. Diabetes. 2000;49:1714-1723. doi: 10.2337/diabetes.49.10.1714.

[64] Acton S, Rigotti A, Landschulz KT, Identification of scavenger receptor SR-BI as a high density lipoprotein receptor. Science. 1996;271: 518-520.

[65] Krieger M. Scavenger receptor class B type I is a multiligand HDL receptor that influences diverse physiologic systems. Journal of Clinical Investigation. 2001;108:793-797.

[66] Endemann G, Stanton L-W, Madden K-S, et al. CD36 is a receptor for oxidized low density lipoprotein. Journal of Biological Chemistry. 1998;268:11811-11816.

[67] Ohgami N, Nagai R, Ikemoto M, et al. Cd36, a member of the class b scavenger receptor family, as a receptor for advanced glycation end products. Journal of Biological Chemistry. 2001;276:3195-3202. doi: 10.1074/jbc. M006545200.

[68] Silverstein RL, Febbraio M. CD36, a Scavenger Receptor Involved in Immunity, Metabolism, Angiogenesis, and Behavior. Science Signaling. 2009;2: re3. doi: 10.1126/scisignal.272re3.

[69] Tamura Y, Adachi H, Osuga J, et al. FEEL-1 and FEEL- 2 are endocytic receptors for advanced glycation end products. Journal of Biological Chemistry. 2003;278:12613-12617

[70] Farquhar MG, Saito A, Kerjaschki D, et al. The Heymann nephritis antigenic complex: Megalin (gp330) and RAP. Journal of the American Society of Nephrology. 1995;6:35-47.

[71] Saito A, Kazama JJ, Iino N, et al. Bioengineered implantation of megalin-expressing cells: A potential intracorporeal therapeutic model for uremic toxin protein clearance in renal failure. Journal of the American Society of Nephrology. 2003;14:2025-2032.

[72] McRobert EA, Gallicchio M, Jerums G, et al. The amino- terminal domains of the ezrin, radixin, and moesin (ERM) proteins bind advanced glycation end products, an interaction that may play a role in the development of diabetic complications. J Biol Chem. 2003; 278: 25783- 25789.

[73] Wang Q, Fan A, Yuan Y, et al. Role of Moesin in Advanced Glycation End Products-Induced Angiogenesis of Human Umbilical Vein Endothelial Cells. Scientific Reports. 2016;6:22749. doi: 10.1038/srep22749.

[74] Schmidt AM, Vianna M, Gerlach M, et al. Isolation and characterization of two binding proteins for advanced glycosylation end products from bovine lung which are present on the endothelial cell surface. Journal of Biological Chemistry. 1992;267:14987-14997

[75] Kierdorf K, Fritz G. RAGE regulation and signaling in inflammation and beyond. Journal of Leukocyte Biology. 2013;94:55-68

[76] Schmidt AM, Yan SD, Yan SF, Stern DM: The multiligand receptor RAGE as a progression factor amplifying immune and inflammatory responses. J Clin Invest 2001, 108:949-955.

[77] Bierhaus A, Humpert PM, Morcos M, et al. Understanding RAGE, the receptor for advanced glycation end products. Journal of Molecular Medicine. 2005;83:876-886.

[78] Katakami N, Matsushima M, Kaneto H, et al. Endogenous secretory RAGE but not soluble RAGE is associated with carotid atherosclerosis in type 1 diabetes patients. Diabetes & Vascular Disease Research. 2008;5:190-197.

[79] Dalton TP, Shertzer HG, Puga A. Regulation of gene expression by reactive oxygen. Annual Review of Pharmacology and Toxicology1999;39:67-101. doi: 10.1146/annurev.pharmtox.39.1.67.

[80] Chen J, Jing J, Yu S, et al. Advanced glycation endproducts induce apoptosis of endothelial progenitor cells by activating receptor RAGE and NADPH oxidase/JNK signaling axis. American Journal of Translational Research. 2016;8:2169-2178.

[81] Chang JS, Wendt T, Qu W, et al. Oxygen deprivation triggers upregulation of early growth response-1 by the receptor for advanced glycation end products. Circulation Research. 2008;102:905-913. doi: 10.1161/CIRCRESAHA.107.165308.

[82] Sakurai S, Yonekura H, Yamamoto Y, et al. The AGE- RAGE system and diabetic nephropathy. Journal of the American Society of Nephrology. 2003;14:S259-S263.

[83] Raucci A, Cugusi S, Antonelli A, et al. A soluble form of the receptor for advanced glycation endproducts (RAGE) is produced by proteolytic cleavage of the membrane-bound form by the sheddase a disintegrin and metalloprotease 10 (ADAM10). FASEB J. 2008;22:3716-3727.

[84] Yonekura H, Yamamoto Y, Sakurai S, et al. Novel splice variants of the receptor for advanced glycation end- products expressed in human vascular endothelial cells and pericytes, and their putative roles in diabetes-induced vascular injury. Biochemical Journal. 2003;370:1097-1109.

[85] Koyama H, Shoji T, Yokoyama H, et al. Plasma level of endogenous secretory RAGE is associated with components of the metabolic syndrome and atherosclerosis. Arteriosclerosis Thrombosis and Vascular Biology. 2005;25:2587-2593.

[86] Katakami N, Matsuhisa M, Kaneto H, et al. Serum endogenous secretory RAGE levels are inversely associated with carotid IMT in type 2 diabetic patients. Atherosclerosis. 2007;190:22-23

[87] Lu L, Pu LJ, Zhang Q, et al. Increased glycated albumin and decreased esRAGE levels are related to angiographic severity and extent of coronary artery disease in patients with type 2 diabetes. Atherosclerosis. 2009; 206: 540-545.

[88] Hancock JT, Desikan R, Neill SJ. Role of Reactive Oxygen Species in Cell Signaling Pathways. Biochemical and Biomedical Aspects of Oxidative Modification. 2001;29:345-350.

[89] Fedorova M, Bollineni RC, Hoffmann R. Protein carbonylation as a major hallmark of oxidative damage: update of analytical strategies. Mass Spectrometry Reviews. 2014;33:79-97. doi: 10.1002/mas.21381.

[90] Rodríguez-García A, García-Vicente R, Morales ML. Protein Carbonylation and Lipid Peroxidation in Hematological Malignancies. Antioxidants (Basel). 2020;9:1212. doi: 10.3390/antiox9121212.

[91] Rowan S, Bejarano E, Taylor A. Mechanistic targeting of advanced glycation end-products in age-related

diseases. Biochimica et Biophysica Acta - Molecular Basis of Diseas. 2018;1864:3631-3643. doi: 10.1016/j.bbadis.2018.08.036.

[92] Li Z, Xu X, Leng X, et al. Roles of reactive oxygen species in cell signaling pathways and immune responses to viral infections. Archives of Virology. 2017;162:603-610. doi: 10.1007/s00705-016-3130-2.

[93] Shadel GS, Horvath TL. Mitochondrial ROS signaling in organismal homeostasis. Cell. 2015;163:560-569. https://doi.org/10.1016/j.cell.2015.10.001

[94] Kusunoki H, Miyata S, Ohara T, et al. Relation between serum 3-deoxyglucosone and development of diabetic microangiopathy Diabetes Care. 2003;26:1889-1894. doi: 10.2337/diacare.26.6.1889.

[95] Wan TT, Li XF, Sun YM, et al. Recent advances in understanding the biochemical and molecular mechanism of diabetic retinopathy. Biomedicine & Pharmacotherapy. 2015;74:145-147

[96] Low Wang CC, Hess CN, Hiatt WR, et al. Clinical Update: Cardiovascular Disease in Diabetes Mellitus: Atherosclerotic Cardiovascular Disease and Heart Failure in Type 2 Diabetes Mellitus - Mechanisms, Management, and Clinical Considerations. Circulation. 2016;133:2459-502. doi: 10.1161/CIRCULATIONAHA.116.022194.

[97] Forbes JM, Cooper ME. Mechanisms of diabetic complications. Physiological Reviews. 2013;93(1):137-88.

[98] Madonna R, Balistreri CR, Geng YJ, et al. Diabetic microangiopathy: pathogenetic insights and novel therapeutic approaches. Vascular Pharmacology. 2017;90:1-7

[99] Stitt AW. AGEs and diabetic retinopathy. Investigative

Ophthalmology & Visual Science. 2010;51:4867-4874. doi: 10.1167/iovs.10-5881.

[100] Kaji Y, Usui T, Oshika T, et al. Advanced glycation end products in diabetic corneas. Investigative Ophthalmology & Visual Science. 2000;41:362-8.

[101] Kaji Y, Usui T, Ishida S, et al. Inhibition of diabetic leukostasis and blood-retinal barrier breakdown with a soluble form of a receptor for advanced glycation end products. Investigative Ophthalmology & Visual Science. 2007;48:858-65. doi: 10.1167/iovs.06-0495.

[102] Barile GR, Pachydaki SI, Tari SR, et al. The RAGE axis in early diabetic retinopathy. Investigative Ophthalmology & Visual Science. 2005;46:2916-24. doi: 10.1167/iovs.04-1409.

[103] American Diabetes Association. Position Statement: Standards of Medical Care in Diabetes - 2016. Diabetes Care. 2016;39:S1-S112.

[104] Tanji N, Markowitz GS, Fu C, et al. Expression of advanced glycation end products and their cellular receptor RAGE in diabetic nephropathy and nondiabetic renal disease. Journal of the American Society of Nephrology. 2000;11:1656-1666.

[105] Inagi R, Yamamoto Y, Nangaku M, et al. A severe diabetic nephropathy model with early development of nodule-like lesions induced by megsin overexpression in RAGE/iNOS transgenic mice. Diabetes. 2006;55:356-366. doi: 10.2337/diabetes.55.02.06.db05-0702.

[106] Vlassara H, Striker LJ, Teichberg S, et al. Advanced glycation end products induce glomerular sclerosis and albuminuria in normal rats. Proceedings of the National Academy of Sciences of

the United States of America.1994;91: 11704-11708.

[107] Iturralde P, Bayes de Luna A, Guindo J. Pharmacologic treatment of ventricular arrhythmias. General considerations and methods for evaluating its effectiveness. Archivos de Cardiología de México. 1987;57: 73-76.

[108] Kilhovd BK, Giardino I, Torjesen PA, et al. Increased serum levels of the specific AGE-compound methylglyoxal-derived hydroimidazolone in patients with type 2 diabetes. Metabolism 2003;52:163-167.

[109] Tan AL, Forbes JM, Cooper ME. AGE, RAGE, and ROS in diabetic nephropathy. Seminars in Nephrology. 2007; 27: 130-143.

[110] Yamamoto Y, Kato I, Doi T, et al. Development and prevention of advanced diabetic nephropathy in RAGE-overexpressing mice. Journal of Clinical Investigation. 2001;108: 261-268.

[111] Reiniger N, Lau K, McCalla D, et al. Deletion of the receptor for advanced glycation end products reduces glomerulosclerosis and preserves renal function in the diabetic OVE26 mouse. Diabetes. 2010;59:2043-2054. doi: 10.2337/db09-1766.

[112] Sugimoto K, Nishizawa Y, Horiuchi S. Localization in human diabetic peripheral nerve of Ne-carboxymethyllysine-protein adducts, an advanced glycation endproduct. Diabetologia. 1997;40:1380-1387

[113] Williams SK, Howarth NL, Devenny JJ, Structural and functional consequences of increased tubulin glycosylation in diabetes mellitus. Proceedings of the National Academy of Sciences of the United States of America. 1982;79:6546-6550. doi: 10.1073/pnas.79.21.6546.

[114] Bierhaus A, Haslbeck KM, Humpert PM. Loss of pain perception in diabetes is dependent on a receptor of the immunoglobulin superfamily. Journal of Clinical Investigation. 2004;114:1741-1751. doi: 10.1172/ JCI18058.

[115] Kihara M, Schmelzer JD, Poduslo JF, Aminoguanidine effects on nerve blood flow, vascular permeability, electrophysiology, and oxygen free radicals. Curran GL, Nickander KK, Low PA. Proc Natl Acad Sci U S A. 1991 Jul 15;88(14):6107-11. doi: 10.1073/ pnas.88.14.6107.

[116] Xu S, Bao W, Men X, et al. Interleukin-10 Protects Schwann Cells against Advanced Glycation End Products-Induced Apoptosis via NF-κB Suppression. Experimental and Clinical Endocrinology & Diabetes. 2020;128:89-96. doi: 10.1055/a-0826-4374.

[117] Yamagishi S, Imaizumi T. Diabetic vascular complications: pathophysiology, biochemical basis and potential therapeutic strategy. Current Pharmaceutical Design. 2005;11:2279-2299. doi: 10.2174/1381612054367300.

[118] Simons M. Angiogenesis: where do we stand now? Circulation. 2005;111:1556-1566. doi: 10.1161/01. CIR.0000159345.00591.8F.

[119] Bucciarelli LG, Wendt T, Qu W, et al. RAGE blockade stabilizes established atherosclerosis in diabetic apolipoprotein E-null mice. Circulation. 2002;106:2827-35. 10.1161/01. CIR.0000039325.03698.36

[120] Yamagishi S, Nakamura K, Matsui T, et al. Receptor for advanced glycation end products (RAGE): a novel therapeutic target for diabetic vascular complication. Current Pharmaceutical Design. 2008;14:487-495. doi: 10.2174/138161208783597416.

[121] Egaña-Gorroño L, López-Díez R, Yepuri G, wt al, Receptor for Advanced Glycation End Products (RAGE) and Mechanisms and Therapeutic Opportunities in Diabetes and Cardiovascular Disease: Insights From Human Subjects and Animal Models. Frontiers in Cardiovascular Medicine. 2020;7:37. doi: 10.3389/ fcvm.2020.00037. eCollection 2020.

[122] Chang JS, Wendt T, Qu W, et al. Oxygen deprivation triggers upregulation of early growth response-1 by the receptor for advanced glycation end products. Circulation Research. 2008;102:905-913. doi: 10.1161/ CIRCRESAHA.107.165308.

[123] Nasu R, Seki K, Nara M, et al. Effect of a high-fat diet on diabetic mother rats and their offspring through three generations. Endocrine Journal.2007;54:563-569. doi: 10.1507/ endocrj.k06-175.

[124] Nasu-Kawaharada R, Nakamura A, Kakarla SK, et al. A maternal diet rich in fish oil may improve cardiac Akt-related signaling in the offspring of diabetic mother rats. Nutrition. 2013 Apr;29(4):688-92. doi: 10.1016/j. nut.2012.11.017.

[125] Kawaharada R, Masuda H, Chen Z, et al. Intrauterine hyperglycemia-induced inflammatory signalling via the receptor for advanced glycation end products in the cardiac muscle of the infants of diabetic mother rats. European Journal of Nutrition. 2018;57:2701-2712. doi: 10.1007/ s00394-017-1536-6.

[126] Agarwal P, Morriseau TS, Kereliuk SM, et al. Maternal obesity, diabetes during pregnancy and epigenetic mechanisms that influence the developmental origins of cardiometabolic disease in the offspring. Critical Reviews in Clinical Laboratory Sciences. 2018;55:71-101. doi: 10.1080/10408363.2017.1422109.

[127] Barker DJ. The fetal and infant origins of adult disease. British Medical Journal. 1990;30:1111.

[128] Barker DJ. The origins of the developmental origins theory. Journal of Internal Medicine. 2007;261:412-417.

[129] Barker DJ, Gluckman, PD, Godfrey KM, et al. Fetal nutrition and cardiovascular disease in adult life. Lancet. 199;341:938-941.

[130] Barker DJ, Eriksson JG, Forsén T, et al. The fetal origins of adult disease. International Journal of Epidemiology. 2002;31:1235-1239.

[131] Hanson MA, Gluckman PD. Early Developmental Conditioning of Later Health and Disease: Physiology or Pathophysiology? Physiological Reviews. 2014;94:1027-1076.

[132] Zhao W, Tilton RG, Corbett JA, et al. Experimental allergic encephalomyelitis in the rat is inhibited by aminoguanidine, an inhibitor of nitric oxide synthase. Journal of Neuroimmunology. 1996;64:123-133.

[133] Price DL, Rhett PM, Thorpe SR, et al. Chelating activity of advanced glycation end-product inhibitors. Journal of Biological Chemistry. 2001;276:48967-48972. doi: 10.1074/jbc. M108196200.

[134] Thornalley PJ. Use of aminoguanidine (Pimagedine) to prevent the formation of advanced glycation endproducts. Archives of Biochemistry and Biophysics. 2003;419: 31-40.

[135] Nakamura S, Makita Z, Ishikawa S, et al. Progression of nephropathy in spontaneous diabetic rats is prevented by OPB-9195, a novel inhibitor of advanced glycation. Diabetes. 1997;46:895-899. doi: 10.2337/ diab.46.5.895.

[136] Figarola JL, Scott S, Loera S, et al. LR-90 a new advanced glycation endproduct inhibitor prevents progression of diabetic nephropathy in streptozotocin-diabetic rats. Diabetologia. 2003;46:1140-1152. doi: 10.1007/s00125-003-1162-0.

[137] Forbes JM, Soulis T, Thallas V, Pangiotopoulos S, Long D, Vasan S, Wagle D, Jerums G, Cooper M: Renoprotective effects of a novel inhibitor of advanced glycation. Diabetologia. 2001;44 :108 –114

[138] Wilkinson-Berka JL, Kelly DJ, Koerner SM, et al. ALT-946 and aminoguanidine, inhibitors of advanced glycation, improve severe nephropathy in the diabetic transgenic (mREN-2)27 rat. Diabetes. 2002;51:3283-3289. doi: 10.2337/diabetes.51.11.3283.

[139] Nagaraj RH, Sarkar P, Mally A, et al. Effect of pyridoxamine on chemical modification of proteins by carbonyls in diabetic rats: characterization of a major product from the reaction of pyridoxamine and methylglyoxal. Archives of Biochemistry and Biophysics. 2002;402:110-119. doi: 10.1016/S0003-9861(02)00067-X.

[140] Thomas MC, Baynes JW, Thorpe SR, et al. The role of AGEs and AGE inhibitors in diabetic cardiovascular disease. Current Drug Targets. 2005;6:453-474. doi: 10.2174/1389450054021873.

[141] Hammes HP, Du X, Edelstein D, et al. Benfotiamine blocks three major pathways of hyperglycemic damage and prevents experimental diabetic retinopathy. Nature Medicine. 2003;9:294-299. doi: 10.1038/nm834.

[142] Miyata T, van Ypersele de Strihou C, Ueda Y, et al. Angiotensin II receptor antagonists and angiotensin-converting enzyme inhibitors lower in vitro the formation of advanced glycation end products: biochemical mechanisms. Journal of the American Society of Nephrology. 2002;13:2478-87. doi: 10.1097/01.asn.0000032418. 67267.f2.

[143] Beisswenger P, Ruggiero-Lopez D. Metformin inhibition of glycation processes. Diabetes & Metabolism, 2003;29:6S95-103. doi: 10.1016/s1262-3636(03)72793-1.

[144] Vasan S, Zhang X, Zhang X, et al. An agent cleaving glucose-derived protein crosslinks in vitro and in vivo. Nature. 1996;382:275-278. doi: 10.1038/382275a0.

[145] Kass DA, Shapiro EP, Kawaguchi M, et al. Improved arterial compliance by a novel advanced glycation end-product crosslink breaker. Circulation. 2001;104:1464-1470. doi: 10.1161/hc3801.097806.

[146] Nagamatsu R, Mitsuhashi S, Shigetomi K, et al. Cleavage of α-dicarbonyl compounds by terpene hydroperoxide. Bioscience, Biotechnology, and Biochemistry.2012;76:1904-1908. doi: 10.1271/bbb.120378.

[147] Bongarzone S, Savickas V, Luzi F, et al. Targeting the Receptor for Advanced Glycation Endproducts (RAGE): A Medicinal Chemistry Perspective. Journal of Medicinal Chemistry. 2017;60:7213-7232. doi: 10.1021/acs. jmedchem.7b00058.

Section 3

Glycoproteomics

Improving the Study of Protein Glycosylation with New Tools for Glycopeptide Enrichment

Minyong Chen, Steven J. Dupard, Colleen M. McClung,

Cristian I. Ruse, Mehul B. Ganatra, Saulius Vainauskas,

Christopher H. Taron and James C. Samuelson

Abstract

High confidence methods are needed for determining the glycosylation profiles of complex biological samples as well as recombinant therapeutic proteins. A common glycan analysis workflow involves liberation of N-glycans from glycoproteins with PNGase F or O-glycans by hydrazinolysis prior to their analysis. This method is limited in that it does not permit determination of glycan attachment sites. Alternative proteomics-based workflows are emerging that utilize site-specific proteolysis to generate peptide mixtures followed by selective enrichment strategies to isolate glycopeptides. Methods designed for the analysis of complex samples can yield a comprehensive snapshot of individual glycans species, the site of attachment of each individual glycan and the identity of the respective protein in many cases. This chapter will highlight advancements in enzymes that digest glycoproteins into distinct fragments and new strategies to enrich specific glycopeptides.

Keywords: glycoproteomics, glycopeptide enrichment, lectin, Fbs1, BGL, O-glycoprotease, alpha-lytic protease

1. Introduction

Protein glycosylation is a post-translational carbohydrate ('glycan') modification of eukaryotic proteins that may affect their folding, stability, localization and biological function. Glycan profiles differ from cell type to cell type and are known to be altered in carcinogenesis [1, 2], inflammation [3], and Alzheimer's disease [4] etc. Importantly, circulating plasma proteins may serve as biomarkers as altered glycosylation profiles may signal specific types of disease. Glycosylation may be assessed on a global level by isolating total protein from tissue, cells or serum followed by liberation of glycans and analysis by reliable methods such as ultra performance liquid chromatography (UPLC) coupled to mass spectrometry (MS). Profiling liberated glycans is useful in some cases, but site of attachment (glycosite) and the glycan structure at each glycosite is more valuable. This "total picture" is attainable when performing liquid chromatography (LC) with tandem mass spectrometry (LC–MS/MS) at the glycopeptide level. Furthermore, it is critically important to be able to strictly characterize therapeutic proteins to ensure reproducible glycosylation, safety and efficacy as the absence, presence or type of glycan is known to dictate the efficacy of some therapeutic molecules [5].

Glycans are attached to certain asparagine residues (N-linked glycans) or serine/threonine residues (O-linked glycans). N-linked glycosylation occurs at the Asn-X-Ser/Thr/Cys (where X is not proline) consensus sequence on proteins that pass through the eukaryotic secretory pathway. There are three structural classes of N-glycan that share a common trimannosyl chitobiose core motif (Man3GlcNAc2) (**Figure 1a**). This core may be further variably decorated with mannose, fucose, galactose, sialic acid, N-acetylgalactosamine (GalNAc) and N-acetylglucosamine (GlcNAc). In contrast to N-glycans, O-glycans are appended to the hydroxyl oxygen of Ser or Thr residues with no strong consensus sequence defining a glycosite. There are eight structural classes of O-glycans that are defined by core di- or tri-saccharides that occupy a glycosite (**Figure 1b**). Each of these cores can be further elaborated with other sugars yielding a large variety of possible O-glycan structures. Over 10% of secreted human proteins carry some form of O-glycan modification. A second form of O-glycosylation occurs on nuclear and cytoplasmic proteins, where a single β-linked GlcNAc is attached to Ser or Thr residues. β-O-GlcNAcylation is an essential, dynamic modification that is important in cell signaling and differentiation [6]. Finally, chemical groups (*e.g.*, sulfate, phosphate, acetate, methyl, etc.) may also occur at various positions on certain N- and O-glycan sugars [7].

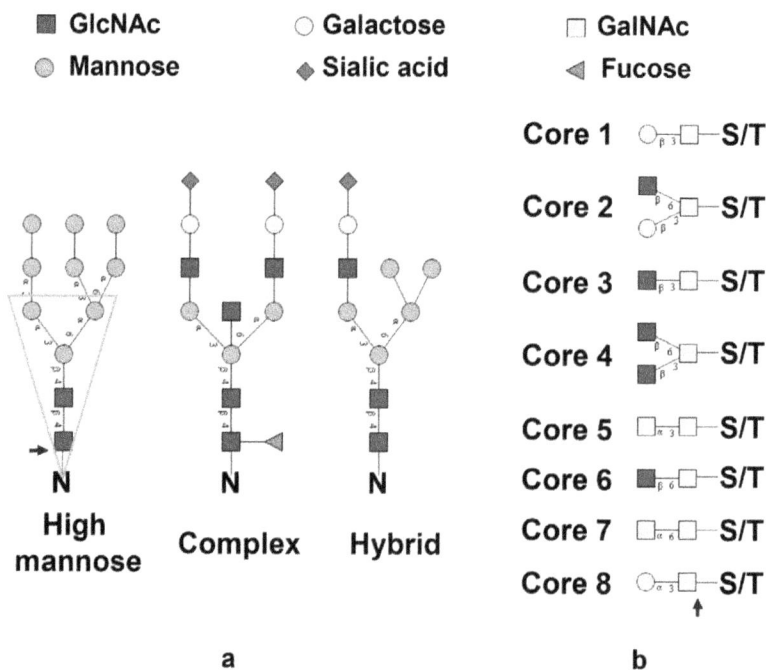

Figure 1.
Basic structures of N-glycans (a) and core structures of O-glycans (b). a, N-glycans can be categorized into three basic types: High mannose, Complex and Hybrid. The core structure (Man3GlcNAc2) of N-glycans is indicated by the orange triangle. A GlcNAc residue can attach to a β-mannose of the N-linked glycan core, resulting in a bisecting N-glycan (illustrated in Complex N-glycan). The reducing end GlcNAc (indicated by an arrow) of N-glycans can also be modified with a fucose (illustrated in Complex N-glycan). A N-glycan modifies a peptide via its reducing end GlcNAc attaching to Asparagine (N) within the peptide. b, eight core structures of O-glycans. O-glycan starts with a GalNAc (reducing end, indicated by an arrow), and further modifications can be added to the non-reducing end of the core structures. In O-glycopeptides, O-glycans are attached to the hydroxyl group of Serine (S) or Threonine (T) via the reducing end GalNAc.

Protein glycosylation is remarkable in its structural complexity. This trait reflects the way in which glycans are synthesized and transferred to proteins. Glycans are assembled by complex biosynthetic pathways consisting of many different enzymes. Individual monosaccharides become linked together by glycosyltransferases that each have sugar and stereochemical specificity. For example, the elaborate mammalian 14 sugar N-glycan precursor consists of only three types of monosaccharides (Glc3Man9GlcNAc2), yet its assembly requires the coordinated action of 13 different glycosyltransferases. There are over 200 different glycosyltransferases that affect glycan structures in the mammalian glycome [8]. Gene expression of some of these enzymes varies by tissue, cell type, and epigenetic regulation resulting in significant structural variation of glycans. A glycoform is a single protein isoform having a defined glycan present at each glycosite. As such, proteins naturally exist as collections of glycoforms. Additionally, some protein isoforms periodically lack glycan occupancy at a potential glycosite. The complexity of these attributes of glycoproteins underscores the technical challenges associated with deconvoluting any given glycome.

Analysis of glycan structure has been performed several different ways. However, the most common approaches typically utilize one of two strategies: (i) analysis of glycans that have been released from glycoproteins or ii) bottom-up proteomics analysis of peptide/glycopeptide mixtures. Standard N-glycan profiling methods begin with liberation of N-glycans from a glycoprotein with the enzyme PNGase F. Typically, they are then labelled at their reducing ends with a fluorophore, and separated via high/ultra-performance liquid chromatography (H/UPLC) or capillary electrophoresis (CE) with fluorescence detection and optional inline mass detection [9]. Glycan structures are assigned to observed peaks by comparing mobility and mass data to glycan reference databases [10]. Exoglycosidases with precise specificities can be used to further confirm structural assignments [11, 12]. For O-glycans, no enzyme that releases a broad range of elaborated O-glycan structures has been identified. Chemical release of O-glycans via hydrazinolysis can be achieved, but this can damage some released glycans [13, 14]. In addition, released N- and O-glycans may be permethylated and analyzed directly by LC–MS/MS or MALDI-MS [15]. Finally, for both N- and O-glycans, profiling of released glycans provides a catalog of the range of structures present in a sample, but it does not provide information regarding their point of attachment in a protein.

A more data-rich method of glycoprotein analysis uses bottom-up proteomics to analyze peptide/glycopeptide mixtures. In this approach, a glycoprotein is treated with a protease (*e.g.*, trypsin) to generate a pool of peptides that are then analyzed by mass spectrometry (typically LC–MS/MS). Data are processed by computer algorithms with the help of protein and glycan mass reference databases (*e.g.*, Byonic software and O-Pair Search) to generate a peptide map and identify appended glycans. Advantages of this method are that the same workflow can yield information about both N- and O-glycans (and other protein modifications), it identifies glycosites, it can determine both glycan occupancy and the range of glycan structures at each glycosite, and it can be quantitative. This approach (termed the 'multi-attribute method', MAM) is gaining traction in the pharmaceutical industry for monitoring the purity of biologic drugs and is expected to become the industry standard for final product characterization [16]. Despite its benefits, there are still technical challenges facing glycoproteomics analyses. For example, existing proteases (*e.g.*, trypsin) used in proteomics often generate large peptides that may have multiple glycans (especially for O-glycans that tend to be clustered within proteins). These generated glycopeptides can be either too large to detect by MS or it can be difficult to assign glycosites on such peptides with high confidence. Therefore, better approaches are needed to generate glycopeptides. Additionally, glycopeptides represent a small portion of a peptide mixture and often do not ionize well. New methods that address sample complexity through enrichment of specific glycopeptides are emerging.

The field has begun to address these issues through development of new reagents that aid in glycoproteomics. Novel proteases, including those that have specificity for O-glycans, have recently been characterized and validated in glycoproteomics workflows. Additionally, reagents and methods that permit selective enrichment of glycopeptides have been applied to reduce sample complexity. In this chapter, we review advances in glycopeptide generation and enrichment methods that are helping to improve glycopeptide analysis. Additionally, we present an example case study illustrating N-glycopeptide enrichment to address glycan heterogeneity in Wnt signaling.

2. A workflow for intact glycopeptide identification

A common strategy to determine protein glycosylation is shown in **Figure 2**. This process generally involves: (i) protease treatment of protein(s) to generate a peptide/glycopeptide mixture, (ii) glycopeptide enrichment, (iii) analysis of isolated glycopeptides by LC–MS/MS, (iv) computational analysis of mass data

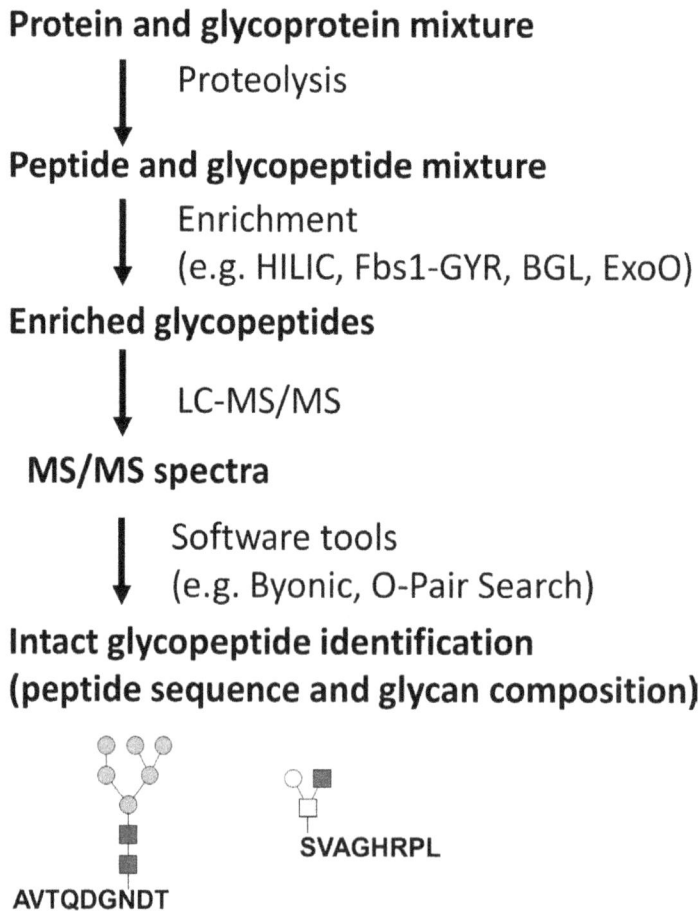

Protein and glycoprotein mixture

↓ Proteolysis

Peptide and glycopeptide mixture

↓ Enrichment
(e.g. HILIC, Fbs1-GYR, BGL, ExoO)

Enriched glycopeptides

↓ LC-MS/MS

MS/MS spectra

↓ Software tools
(e.g. Byonic, O-Pair Search)

**Intact glycopeptide identification
(peptide sequence and glycan composition)**

SVAGHRPL

AVTQDGNDT

Figure 2.
Basic workflow of intact glycopeptide identification.

against proteome and glycan reference databases to yield both the peptide sequence and possible glycan structure for each peptide. Here we review technical challenges and recent advances for each step of glycopeptide analysis.

2.1 Peptide generation

To generate peptides for proteomics analyses, a protein sample is first digested with a protease. The specificity of the protease used can significantly impact the protein coverage obtained by the method. Trypsin, a protease that cleaves after lysine and arginine residues, has been the workhorse of the proteomics field for over two decades. Trypsin generally produces peptides of sufficient length to ionize efficiently in mass spectrometry. However, protein-specific challenges can occur with trypsin, especially with glycoproteins. For example, some proteins naturally lack lysine or arginine residues, have these residues disparately positioned, or have bulky glycans in close proximity that sterically hinder proteolysis. Each of these factors can produce larger peptides that typically do not ionize as well. As such, other proteases with cleavage specificities orthogonal to trypsin are often used to increase proteolytic peptide coverage (**Table 1**) [17]. For example, **Figure 3** shows that α-Lytic Protease can be used alone or in combination with other proteases to yield increased sequence coverage.

A recent advance has been the use of O-glycan-specific proteases (O-endoproteases) for generating O-glycopeptides for analysis. These enzymes recognize and bind to mucin-type O-glycans, then cleave the peptide bond immediately N-terminal to the glycosylated serine or threonine. Used either alone or in series with other proteases like trypsin, glycopeptides are generated that have an O-glycan on their amino-terminal amino acid following cleavage. The first commercial enzyme of this class was the O-endoprotease from *Akkermansia muciniphila* (sold under the trade name OpeRATOR, Genovis AB, Sweden). This enzyme recognizes mammalian O-glycans but it is inhibited by the presence of terminal sialic acids. Accordingly, sialidase treatment is required for efficient performance which results in loss of glycan structural information. Recently, chemical modification of sialic acids has also been shown to improve OpeRATOR function [18]. In contrast, the O-glycoprotease newly available from New England Biolabs, is not inhibited by the presence of sialic acids and it also exhibits a broad specificity towards proteins with mammalian

	Protease	Specificity
N-terminal cleavage	AspN	D (E)
	LysN	K
	LysargiNase	R, K
	O-endoprotease	S/T with O-glycan
C-terminal cleavage	ArgC	R (K)
	GluC	E (D)
	LysC	K
	Trypsin	K, R
	chymotrypsin	F, Y, L, W, M
	Pepsin	Y, F, W
	α-Lytic Protease	T, A, S, V (C, L)

Table 1.
Proteases used in proteomics. Protease specificities are indicated using single letter codes for amino acid residues. Recognition sites that are cleaved at a lower rate are indicated by amino acids bracketed by parentheses.

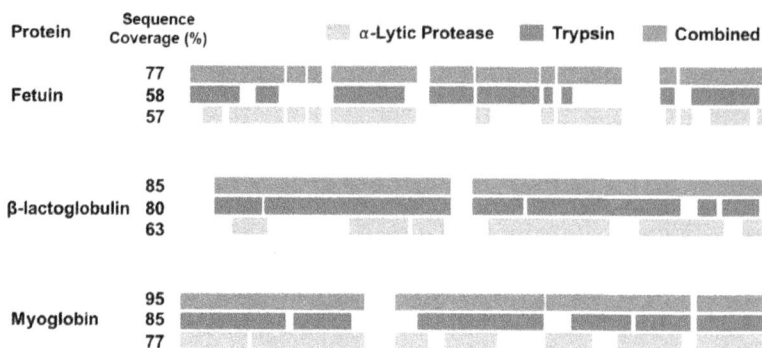

Figure 3.
α-Lytic Protease can be used alone or in combination with other proteases to yield increased sequence coverage. Comparison of sequence coverage for three protein standards after parallel digestion using Trypsin (blue) or a-Lytic Protease (gold). The combined data set (grey) results in overlapping peptides and increased sequence coverage. (Reprinted by permission from New England Biolabs. https://www.neb.com/products/ p8113-a-lytic-protease#Product%20Information).

O-glycans. This enzyme recognizes O-glycans ranging in size from a minimal GalNAc-α-Ser/Thr structure to larger mucin-type O-glycans bearing branches and sialic acids. This specificity negates the need for sialidase treatment or chemical modification prior to O-glycopeptide generation. Resulting O-glycopeptides can be mapped to identify the protein of origin, the position of O-glycosites, and the range of O-glycan structures present at any given glycosite in a single experiment.

2.2 Glycopeptide enrichment methods

Glycopeptides are typically in low abundance compared to aglycosylated peptides in a peptide mixture. Additionally, it is well-established that ionization of glycopeptides is often weaker compared to aglycosylated peptides during MS analyses [19]. This results in aglycosylated peptide signals often dominating MS experiments. Therefore, enrichment of glycopeptides prior to sample analysis has been a growing trend to improve intact glycopeptide identification. Several enrichment schemes that vary in their rationales have been described. These approaches range from general methods (enrichment of both N- and O-linked glycopeptides) to newer glycan class-specific approaches that selectively enrich for either N- or O-linked glycopeptides. Several approaches are summarized here.

2.2.1 Hydrophilic interaction liquid chromatography (HILIC)

HILIC has been widely used for glycopeptide enrichment. It is based on the interaction between the hydrophilic glycan moiety of a glycopeptide and the polar stationary phase in the non-polar mobile phase (typically acetonitrile). Many HILIC materials have been developed, however zwitterionic HILIC (ZIC-HILIC) enrichment is generally the most useful due to higher loading capacity and broader specificity. HILIC does not discriminate between O-linked and N-linked glycopeptides and hydrophilic non-glycosylated peptides may co-elute [20]. Thus, for more complete glycopeptide enrichment, HILIC may require a complementary chromatography fractionation step [21].

2.2.2 Boronic acid

One method utilizes boronic acid presented on a solid support to react with cis-diol-containing saccharides or polyols to form five- or six-membered cyclic esters. This property has been used to capture glycoproteins and glycopeptides [22]. Importantly, the covalent linkage is reversible at acidic pH which results in release of intact glycopeptides [20]. The interaction between boronic acid and sugars is relatively weak but newly characterized derivatives show promise for enrichment of low-abundance glycopeptides [19]. A final consideration is that boronic acid enrichment does not discriminate between N- and O-linked glycopeptides.

2.2.3 Metal affinity chromatography

This method exploits the ability of negatively charged sialylated glycans to coordinate with titanium, zirconium or silver [23]. However, metal ion affinity chromatography is not strictly selective for sialylated glycans as negatively charged phosphopeptides or acidic peptides may compete for binding. Additionally, the method does not discriminate between N- and O-linked glycopeptides.

2.2.4 Hydrazide chemistry

Hydrazide chemistry has been widely used for glycosite characterization. Cis-diols within glycans of glycopeptides may be oxidized to aldehydes (using periodate oxidation) forming a non-reversible covalent bond with hydrazide immobilized on a bead. PNGase F is then used to release the formerly N-linked glycosylated peptides to enable N-glycosite determination using MS [24]. Although more commonly used for N-glycosite determination, this chemistry can also be used to enrich glycopeptides having sialylated glycans. In this method, mild periodate treatment selectively oxidizes sialic acids thus enabling capture of sialyated N- and O-glycopeptides on hydrazide beads. The intact glycopeptides can then be selectively released by acid hydrolysis and analyzed by MS [25].

2.2.5 Enzyme-mediated O-glycopeptide enrichment

O-glycopeptides may be enriched using an enzyme-based workflow termed "EXoO" (extraction of O-linked glycopeptides) [26]. This method is enabled by the availability of O-endoproteases (described above). The workflow (**Figure 4**) involves digestion of a protein/biological sample with a standard protease such as trypsin to generate a peptide mixture. The peptides are conjugated to a solid support via the terminal NH_2 group on each peptide (*e.g.*, Aminolink™ beads, ThermoFisher). An O-endoprotease is used to specifically release O-glycopeptides from the beads. Efficiency of the method is dependent on the specificity of the O-endoprotease. This approach may be practiced with OpeRATOR (Genovis) following chemical modification of sialic acids [18] or with O-glycoprotease (New England Biolabs) which cleaves without pre-treatment to remove or modify sialic acids.

2.2.6 Native lectin-mediated glycopeptide enrichment

Lectins are non-catalytic proteins that bind to carbohydrates. Lectins have been used in a variety of glycan, glycoprotein and glycopeptide enrichment strategies. A common approach utilizes broad-specificity bead-immobilized lectins to capture a wide spectrum of glycopeptides. For example, the lectins Concanavalin A (ConA) and wheat germ agglutinin (WGA) bind to high mannose structures and GlcNAc or sialic acid residues, respectively. Each has been used to isolate N-glycopeptides from peptide mixtures [27]. However, WGA does not exclusively bind to N-glycopeptides as it also binds O-β-GlcNAc found on intracellular proteins [28]. Similar strategies

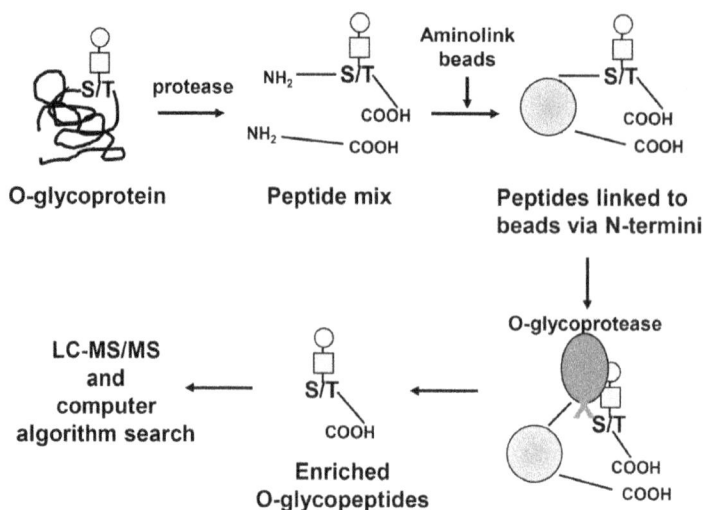

Figure 4.
Basic workflow of O-glycopeptide enrichment by O-glycoprotease.

have been applied to O-glycopeptide enrichment. For example, the lectins Jacalin and *Vicia villosa* agglutinin (VVA) bind to O-linked Gal(β-1,3)GalNAc and α- or β- linked terminal N-acetylgalactosamine, respectively [29, 30].

Lectin-based enrichment strategies have some limitations due to their natural properties. First, most lectins bind their substrates rather weakly (Kd of ~10 mM to 1 µM) [31]. Additionally, limitations in a lectin's specificity can introduce bias into an enrichment scheme. Strategies employing multiple lectins (multi-lectin affinity chromatography, M-LAC) have successfully increased glycopeptide recovery and coverage but do not completely solve the problem of lectin specificity bias [32]. To improve the performance of lectins in glycopeptide enrichment strategies, today's advanced capabilities for cloning and recombinant expression of lectins allows for mutagenesis and selection of lectins with improved binding properties.

2.2.7 Engineered lectins for N- and O-glycopeptide enrichment

The use of structure-guided protein engineering techniques has been used to create lectins with enhanced utility for glycopeptide enrichment. One area of interest has been to engineer binding proteins that can stratify a peptide mixture into different classes of glycopeptides (*e.g.*, N-glycopeptides or O-glycopeptides). Here we summarize recent progress in creating such reagents.

An ideal lectin for N-glycopeptide enrichment would bind to a structurally invariable portion of the N-glycan structure. A common trimannosyl chitobiose (Man3GlcNAc2) core glycan is a common feature of all N-glycans (**Figure 1a**). The human Fbs1 protein specifically recognizes this core motif [33, 34]. Fbs1 participates in glycoprotein quality control within the endoplasmic-reticulum-associated degradation (ERAD) system by binding to misfolded glycoproteins that have been retrotranslocated into the cytosol for degradation [35]. As part of the E3 ubiquitin complex, Fbs1 mediates ubiquitination and degradation of glycoproteins by the proteosome [33, 34]. Wild-type (wt) Fbs1 preferentially binds to high mannose N-glycans with sub-micromolar binding affinity (Kd of 0.1–0.2 µM) and only weakly binds to complex N-glycans having terminal sialic acids [36]. To adapt Fbs1 for use as a universal N-glycan/N-glycopeptide binding reagent, Fbs1 variants with greater tolerance for the presence of sialic acids were engineered using a novel plasmid display strategy where library variants were enriched for their ability to bind immobilized fetuin [37]. An Fbs1 variant (termed Fbs1-GYR) containing S155G, F173Y and E174R substitutions was identified that efficiently binds to both high mannose N-glycans and complex N-glycans (**Figure 5**). Fbs1-GYR is unhindered by sialic acid and core fucose substitution, but does not bind to N-glycans bearing bisecting GlcNAc.

Fbs1-GYR is an efficient and substantially unbiased N-glycopeptide enrichment reagent. It enabled a deep characterization of the human serum N-glycoproteome [37] where Fbs1-GYR enrichment outperformed enrichment by the native lectin mixture of WGA, ConA and RCA$_{120}$ (WCR). Fbs1-GYR enrichment enabled identification of 2.2-fold more N-glycopeptides: an average of 2,142 N-glycopeptide spectra with Fbs1-GYR whereas enrichment with the WCR lectin mixture yielded an average of 965 N-glycopeptide spectra when the same amount of sample was analyzed by MS [37]. Fbs1-GYR mediated enrichment may be performed by using the N-glyco FASP method [32] or by using Fbs1-GYR immobilized beads. In the latter case, Fbs1-GYR has been expressed as a fusion to a SNAP-tag which permits covalent conjugation to benzyl-guanine beads [37–39].

A lectin (termed 'BGL') from the North American Kurokawa mushroom (*Boletopsis grisea*) was recently shown to have a specificity suitable for enrichment of a broad range of O-glycan and O-glycopeptide structures [40]. BGL is a member of the fungal fruit body lectins (Pfam PF07367) that possess two ligand binding sites, as verified by x-ray crystallography [41, 42]. One site binds to N-glycans possessing outer-arm terminal GlcNAc and the other to O-glycans bearing the TF-antigen disaccharide Galβ1,3GalNAc [40]. Ganatra *et al.* used structure-guided mutagenesis to generate single ligand binding site BGL variants [40]. One mutant BGL protein (R103Y) lost the ability to bind N-glycans with a terminal GlcNAc but retained the ability to bind O-glycans bearing the Galβ1,3GalNAc epitope. Both the R103Y BGL variant and *wt* BGL were shown to specifically isolate O-glycopeptides from proteolyzed fetuin, a peptide mixture that contains N-, O- and aglycosylated peptides [40]. As the R103Y BGL variant does not bind to N-glycans, it shows promise as a selective O-glycan/O-glycopeptide enrichment reagent (**Figure 6**). It is plausible that BGL (R103Y) and Fbs1-GYR could be used in tandem to stratify glycopeptide mixtures into enriched pools of O- or N-glycopeptides, respectively.

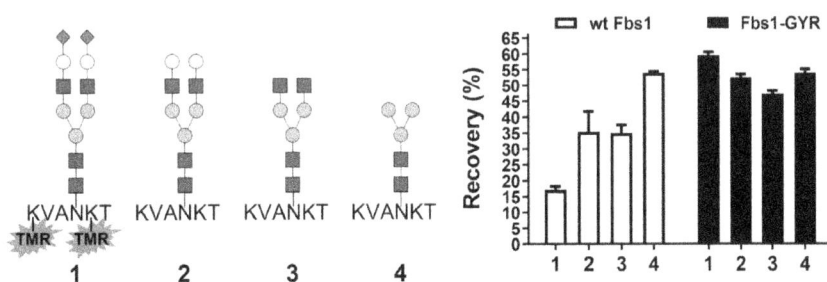

Figure 5.
Fbs1-GYR variant binding to a diverse set of N-glycopeptides is substantially unbiased. Sialylglycopeptide (SGP), an Fbs1 binding substrate, was fluorescently labeled with Tetramethylrhodamine (TMR) at the epsilon-amino group of lysine. For simplicity, TMR is only shown in N-glycopeptide structure 1. N-glycans of SGP-TMR (1) were trimmed with different combinations of exoglycosidases to produce asialo-SGP-TMR (2), SGP-TMR without sialic acids and galactose (3) and SGP-TMR without sialic acids, galactose and GlcNAc (4). The trimmed glycopeptides were then added to binding assays with wt Fbs1 or Fbs1-GYR beads in 50 mM ammonium acetate pH 7.5. The relative binding affinity to wt Fbs1 or Fbs1-GYR is reported as the recovery percentage (TMR fluorescence on beads/input TMR fluorescence). Results represent the mean ± s.e.m. of three replicates. (This figure was originally published within Nature Communications, Volume 8, Article number: 15487 (2017)).

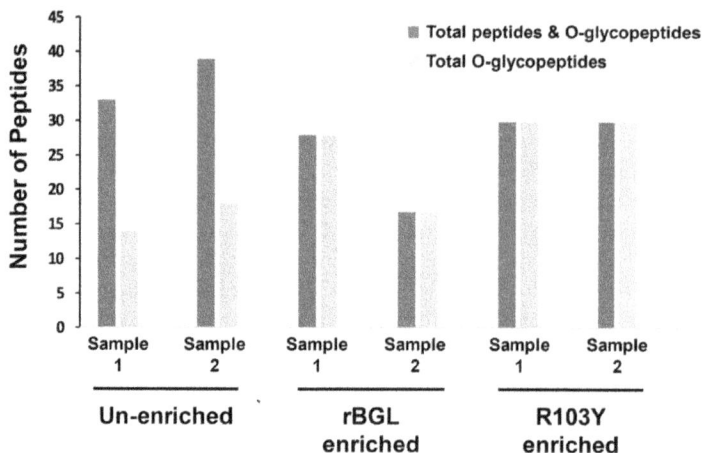

Figure 6.
Enrichment of O-glycosylated peptides/peptiforms from Pronase digested bovine fetuin before or after enrichment with BGL or BGL variant R103Y. Sample 1 and 2 represent replicate samples that were each separately digested with Pronase and subjected to lectin enrichment. Blue bars represent the total number of peptides identified (unglycosylated peptides and O-glycopeptides). Yellow bars represent the number of unique O-glycopeptides/peptiforms identified in each sample. (This figure was originally published within Scientific Reports, Volume 11: Article number: 160 (2021)).

2.3 LC-MS/MS and computer algorithms to search glycopeptides

2.3.1 LC-MS/MS

To identify intact glycopeptides, information of both the peptide backbone and the appended glycan is required. There are four major MS/MS fragmentation methods: collision induced dissociation (CID), electron-capture dissociation (ECD), electron transfer dissociation (ETD), and higher energy collisional dissociation (HCD). CID mainly fragments the peptide backbone, while ECD/ETD is more specific for glycan fragmentation. HCD can fragment both peptide backbone and glycan, and is widely used in intact glycopeptide MS/MS. A combination of different fragmentation methods can improve intact glycopeptide identification.

One recent study reported analysis of more than 5,600 glycopeptides and 1545 N-glycosites [43]. This report implemented a new type of tandem MS fragmentation: activated-ion electron transfer (AI-ETD). The analysis illustrated one of the first studies of glycoproteome profiling with AI-ETD on a quadrupole-Orbitrap-linear ion trap MS system (Orbitrap Fusion Lumos) [44]. Through specialized ion scanning routines, the authors acquired glycopeptide spectra with a higher-energy collision dissociation-product dependent-activated ion electron transfer dissociation (HCD-pd-AI-ETD). This strategy borrows from an established approach in N-glycopeptide analysis, HCD-product ion-triggered-ETD activation where abundant oxonium ions (m/z 204.087, HexNAc) in HCD MS/MS initiate subsequent ETD of the selected precursors [45, 46]. The new method HCD-pd-AI-ETD showed a median of peptide backbone sequence coverage of 89% and a median 78% glycan sequence coverage [44]. These parameters were derived from informatics tools with multiple filtering steps post-analysis. Overall, the filtering strategy aimed to attain no decoy peptide hits within the constraints of below 1% FDR estimations for both AI-ETD and HCD spectra.

2.3.2 Computer algorithms for intact glycopeptide identification

The complicated structure of intact glycopeptides makes the MS/MS spectra extremely complex. Therefore, special computer algorithms have been developed to match the MS/MS spectra to both the peptide sequences and the attached glycan compositions. The algorithms include Byonic [47], GPQuest [48], pGlyco 2.0 [49], and O-pair Search [50]. Amongst these programs, the Byonic search engine provides high sensitivity identification of glycopeptides and allows the use of customized databases for both glycans and proteins. Byonic software identifies glycopeptides to the level of glycan composition and peptide sequence, and it is suitable for both N-glycopeptide and O-glycopeptide searches. A newly published computer algorithm, called O-Pair Search, is specific for O-glycopeptide searches [50]. The authors claim that O-Pair Search can not only greatly reduce search times (up to more than 2,000-fold) compared to a Byonic search, but it can also generate more O-glycopeptide identifications.

3. A case study: application of the Fbs1-GYR enrichment method to study N-glycan heterogeneity in Wnt signaling

The Wnt signaling pathway plays important roles in normal development and in cancer progression [51]. Several enzymes (such as DPAGT1) which are involved in N-glycan biogenesis are regulated by Wnt3a ligand stimulation [52, 53]. Therefore, an N-glycosylation study was performed to reveal potential biomarkers for the detection of Wnt-related cancers. We applied Fbs1-GYR enrichment technology to investigate whether protein N-glycosylation heterogeneity changes upon Wnt3a stimulation in mammalian cells.

Murine recombinant Wnt3a ligand and the Wnt Protein Stabilizer (AMS. bWps) (ASMBio, Cambridge, MA) in combination were able to stimulate canonical Wnt signaling in HEK293 SuperTopFlash STF cells (ATCC CRL-3249) in serum free media. Controls cells (non-Wnt3a stimulated cells) were treated in the same manner but without addition of Wnt3a. A 103.6 ± 3.92 (n = 3) fold change in TopFlash reporter gene expression was observed after 24-hour stimulation with 50 ng/ml Wnt3a and 50 µg/ml of Stabilizer. Note that serum free media was necessary to prevent possible glycoprotein contamination from bovine serum.

Control cells or Wnt3a-stimulated cells were harvested, and total protein was digested with trypsin. N-glycopeptides were enriched with 50 µg Fbs1-GYR puri-fied protein using the N-glyco-FASP method [32] from 200 µg of tryptic peptides prepared from either control cells or Wnt3a-stimulated cells. The enriched N-glycopeptide samples were subjected to LC–MS/MS analysis and N-glycopeptide searching by Byonic as described [37]. 1556 and 1233 N-glycopeptide spectrum matches (N-glyco PSM) were obtained from Wnt3a-stimulated cells and control cells, respectively. (The complete dataset is available upon request from the corre-sponding author). The numbers of peptide spectrum matches (PSM) are suggestive of the relative abundance of the peptides [26]. Thus, the value of N-glyco PSM is used to evaluate and compare protein N-glycosylation in the Wnt3a-stimulated cells and the control cells. Using criteria of at least a two-fold change and a minimum 10 PSM difference between Wnt3a-stimulated and control cells, 17 proteins were identified exhibiting significant changes in N-glycosylation (**Table 2**). Among them, N-glycosylation of 11 proteins (MPRI, AN32B, PON2, SAP, NOMO3, TMED4, FKBP9, ATRN1, LMNB2, ZMAT4, and BASI) showed a significant increase upon Wnt3a stimulation, while N-glycosylation of 6 proteins (MA2B1, PLOD1, MPRD, MOMO1, HEAT1, and ABCAD) was significantly reduced with Wnt3a stimulation. MPRI (Cation-independent M6P receptor) and MPRD (Cation-dependent M6P receptor) are both mannose-6-phosphate (M6P) receptors. However, they display an opposite response to Wnt3a stimulation with regard to N-glycosylation (**Table 2**, **Table 3** highlighted within the dark blue box). The detected N-glycosylation of

	|Uniprot ID| N-glycoprotein identity	Spectral Counts		ratio
		Wnt3a	control	Wnt3a/Control
1	|P11717|MPRI_HUMAN Cation-independent mannose-6-phosphate receptor OS=Homo sapiens GN=IGF2R PE = 1 SV = 3	35	3	11.7
2	|Q92688|AN32B_HUMAN Acidic leucine-rich nuclear phosphoprotein 32 family member B OS=Homo sapiens GN = ANP32B PE = 1 SV = 1	20	6	3.3
3	|Q15165|PON2_HUMAN Serum paraoxonase/ arylesterase 2 OS=Homo sapiens GN=PON2 PE = 1 SV = 3	19	9	2.1
4	|O00754|MA2B1_HUMAN Lysosomal alpha-mannosidase OS=Homo sapiens GN = MAN2B1 PE = 1 SV = 3	3	15	0.2
5	|Q02809|PLOD1_HUMAN Procollagen-lysine,2-oxoglutarate 5-dioxygenase 1 OS=Homo sapiens GN=PLOD1 PE = 1 SV = 2	4	17	0.2
6	|P20645|MPRD_HUMAN Cation-dependent mannose-6-phosphate receptor OS=Homo sapiens GN = M6PR PE = 1 SV = 1	7	22	0.3
7	|P07602|SAP_HUMAN Prosaposin OS=Homo sapiens GN=PSAP PE = 1 SV = 2	63		
8	|P69849|NOMO3_HUMAN Nodal modulator 3 OS=Homo sapiens GN=NOMO3 PE = 2 SV = 2	18		
9	|Q7Z7H5|TMED4_HUMAN Transmembrane emp24 domain-containing protein 4 OS=Homo sapiens GN = TMED4 PE = 1 SV = 1	17		
10	|O95302–3|FKBP9_HUMAN Isoform 3 of Peptidyl-prolyl cis-trans isomerase FKBP9 OS=Homo sapiens GN=FKBP9	13		
11	|Q5VV63|ATRN1_HUMAN Attractin-like protein 1 OS=Homo sapiens GN = ATRNL1 PE = 2 SV = 2	11		
12	|Q03252|LMNB2_HUMAN Lamin-B2 OS=Homo sapiens GN = LMNB2 PE = 1 SV = 3	10		
13	|Q9H898|ZMAT4_HUMAN Zinc finger matrin-type protein 4 OS=Homo sapiens GN = ZMAT4 PE = 2 SV = 1	10		
14	|P35613|BASI_HUMAN Basigin OS=Homo sapiens GN=BSG PE = 1 SV = 2	10		
15	|Q15155|NOMO1_HUMAN Nodal modulator 1 OS=Homo sapiens GN=NOMO1 PE = 1 SV = 5		26	
16	|Q9H583|HEAT1_HUMAN HEAT repeat-containing protein 1 OS=Homo sapiens GN=HEATR1 PE = 1 SV = 3		17	
17	|Q86UQ4|ABCAD_HUMAN ATP-binding cassette sub-family A member 13 OS=Homo sapiens GN = ABCA13 PE = 2 SV = 3		14	

Table 2.
List of 17 proteins with significant differences in overall N-glycosylation upon Wnt3a stimulation.
N-glycosylation is evaluated by spectral counting label-free quantification. The scoring criteria was a minimum
of 10 PSM difference and a fold change minimum of 2 between Wnt3a-stimulated cells and the control cells.

>sp|Uniprot ID|N-glycoprotein name N-glycopeptide; N@denotes N-glycosite N-glycan attaching to the above N-glycopeptide	PSM in Control samples	PSM in Wnt3a samples
>sp|O00754|MA2B1_HUMAN Lysosomal alpha-mannosidase OS=Homo sapiens GN=MAN2B1 PE=1 SV=3	15	3
K.AN@LTWSVK.H	1	
HexNAc(2)Hex(5)	1	
K.LN@QTEPVAGNYYPVNTR.I	2	1
HexNAc(2)Hex(3)Fuc(1)	2	1
R.N@LSAPVTLNLR.D	1	
HexNAc(2)Hex(6)	1	
R.WWHQQTN@ATQEVVR.D	11	1
HexNAc(2)Hex(3)Fuc(1)	11	1
R.YYRTN@HTVMTMGSDFQYENANMWFK.N		1
HexNAc(4)Hex(5)Fuc(1)NeuAc(1)Na(1)		1
>sp|P11717|MPRI_HUMAN Cation-independent mannose-6-phosphate receptor OS=Homo sapiens GN=IGF2R PE=1 SV=3	3	35
K.MN@FTGGDTCHK.V		13
HexNAc(2)Hex(6)		3
HexNAc(2)Hex(7)		5
HexNAc(2)Hex(8)		4
HexNAc(2)Hex(9)		1
K.TN@ITLVCKPGDLESAPVLR.T	1	20
HexNAc(2)Hex(5)		7
HexNAc(2)Hex(5)Fuc(1)		4
HexNAc(2)Hex(6)	1	9
R.N@GSSIVDLSPLIHR.T	2	2
HexNAc(2)Hex(9)	2	2
>sp|P20645|MPRD_HUMAN Cation-dependent mannose-6-phosphate receptor OS=Homo sapiens GN=M6PR PE=1 SV=1	22	7
R.LKPLFN@K.S	22	7
HexNAc(2)Hex(5)	6	
HexNAc(2)Hex(6)	10	2
HexNAc(2)Hex(7)	3	3
HexNAc(2)Hex(8)	3	2
>sp|Q9Y4L1|HYOU1_HUMAN Hypoxia up-regulated protein 1 OS=Homo sapiens GN=HYOU1 PE=1 SV=1	71	119
K.EN@GTDTVQEEEESPAEGSK.D	7	2
HexNAc(2)Hex(5)	4	1
HexNAc(2)Hex(6)	3	1
K.EN@GTDTVQEEEESPAEGSKDEPGEQVELK.E	1	
HexNAc(2)Hex(5)	1	
K.EN@GTDTVQEEEESPAEGSKDEPGEQVELKEEAEAPVEDGSQPPPPEPK.G	7	8
HexNAc(2)Hex(5)	4	3
HexNAc(2)Hex(5)Fuc(1)		1
HexNAc(2)Hex(6)	3	4
K.N@GTR.A	2	9
HexNAc(4)Hex(4)Fuc(1)	1	4
HexNAc(4)Hex(5)Fuc(1)	1	2
HexNAc(4)Hex(5)Fuc(1)NeuAc(2)Na(1)		3
K.VIN@ETWAWK.N	14	20
HexNAc(2)Hex(5)	2	4
HexNAc(2)Hex(6)	6	8
HexNAc(2)Hex(7)	3	3
HexNAc(2)Hex(8)	1	1
HexNAc(2)Hex(9)	2	4
R.AEPPLN@ASASDQGEK.V	19	57
HexNAc(2)Hex(5)	7	11
HexNAc(2)Hex(6)	11	46
HexNAc(2)Hex(8)	1	
R.LSALDNLLN@HSSMFLK.G	2	4
HexNAc(2)Hex(6)	2	3
HexNAc(2)Hex(8)		1
R.VFGSQN@LTTVK.L	19	19
HexNAc(2)Hex(5)	2	
HexNAc(2)Hex(6)	3	4
HexNAc(2)Hex(7)	1	2
HexNAc(2)Hex(8)	5	5
HexNAc(2)Hex(9)	8	8

Table 3.
Comparison of detected N-glycosylation in MA2B1, MPRI, MPRD, and HYOU1 in Wnt3a-stimulated cells and the control cells. The light green rows indicate N-glycoprotein identity. Beneath the protein identity row, N-glycosites are listed in light blue. Beneath each N-glycosite, the respective N-glycan composition is listed. N@ indicates the asparagine with N-glycan modification. PSM numbers of individual N-glycosylation modifications are listed in columns on the right side.

MPRI increases 11.7 fold, while N-glycosylation of MPRD decreases approximately 3-fold after Wnt3a stimulation. The observed N-glycosylation changes may be due to the changes of protein expression level, which deserves further investigation.

The N-glyco PSM of lysosomal alpha-mannosidase (MA2B1(O00754)), is greatly reduced (5-fold) with Wnt3a stimulation. Interestingly, the N-glycosylation change is mainly due to differential glycosylation of N133 of MA2B1(O00754). **Table 3** shows 11 PSM with a fucosylated N-glycan (HexNAc(2)Hex(3)Fuc(1)) were found at position N133 attached to this mannosidase in control cells, but only one PSM was detected in the Wnt3a-stimulated cells (highlighted in the red box). Thus, we speculate that reduced N-glycosylation may affect stability of this mannosidase in the lysosome resulting in altered N-glycosylation of substrate proteins. The extent of N-glycosylation of some enzymes did not differ significantly between Wnt3a-stimulated cells and the control cells. However, modification of a specific N-glycosite did differ significantly. For example, there was no significant fold change with regard to the total numbers of N-glyco PSM of HYOU1, Hypoxia up-regulated protein 1, which were 71 and 119 in control and Wnt3a-stimulated cells, respectively (**Table 3**). However, a 3-fold increase in N-glycosylation at position N931 of HYOU1 was found upon Wnt3a stimulation (57 PSM in Wnt3a-stimulated cells vs. 19 PSM in control cells, **Table 3**, highlighted in the green box). Overall, this study demonstrates that the Fbs1-GYR enrichment method allows for the examination of glycosite heterogeneity of individual cellular proteins, and this study has revealed candidate biomarkers for Wnt-related cancers.

4. Conclusion

Due to the inherent complexity of protein glycosylation, better reagents and workflows are required in order to thoroughly and accurately characterize the glycosylation profile within a sample of interest. Intact glycopeptide identification (glycosite and glycan composition) has emerged as a more effective means to study heterogeneity, to investigate disease biomarkers and to characterize therapeutic proteins. Fortunately, several notable advances have arisen in the last few years. These advances include chemical enrichment strategies, engineered lectins with improved specificity, a greater selection of site-specific proteases, more sophisticated mass spectrometry methods/instruments and finally the development of computer algorithms designed for deconvolution of glycopeptide fragmentation spectra. Although challenges remain, these advances have certainly simplified the study of protein glycosylation.

Acknowledgements

The authors acknowledge New England Biolabs, James V. Ellard and Donald G. Comb for research support.

Author details

Minyong Chen, Steven J. Dupard, Colleen M. McClung, Cristian I. Ruse,
Mehul B. Ganatra, Saulius Vainauskas, Christopher H. Taron
and James C. Samuelson*
New England Biolabs, Inc., Ipswich, MA, USA

*Address all correspondence to: samuelson@neb.com

IntechOpen

References

[1] Christiansen MN, Chik J, Lee L, Anugraham M, Abrahams JL, Packer NH. Cell surface protein glycosylation in cancer. Proteomics. 2014;14(4-5):525-46.

[2] Stuchlová Horynová M, Raška M, Clausen H, Novak J. Aberrant O-glycosylation and anti-glycan antibodies in an autoimmune disease IgA nephropathy and breast adenocarcinoma. Cell Mol Life Sci. 2013;70(5):829-39.

[3] Scott DW, Patel RP. Endothelial heterogeneity and adhesion molecules N-glycosylation: implications in leukocyte trafficking in inflammation. Glycobiology. 2013;23(6):622-33.

[4] Schedin-Weiss S, Winblad B, Tjernberg LO. The role of protein glycosylation in Alzheimer disease. Febs j. 2014;281(1):46-62.

[5] Mizushima T, Yagi H, Takemoto E, Shibata-Koyama M, Isoda Y, Iida S, et al. Structural basis for improved efficacy of therapeutic antibodies on defucosylation of their Fc glycans. Genes Cells. 2011; 16(11):1071-80.

[6] Slawson C, Hart GW. O-GlcNAc signalling: implications for cancer cell biology. Nat Rev Cancer. 2011;11(9): 678-84.

[7] Muthana SM, Campbell CT, Gildersleeve JC. Modifications of glycans: biological significance and therapeutic opportunities. ACS Chem Biol. 2012;7(1):31-43.

[8] Moremen KW, Tiemeyer M, Nairn AV. Vertebrate protein glycosylation: diversity, synthesis and function. Nat Rev Mol Cell Biol. 2012;13(7):448-62.

[9] Duke R, Taron CH. N-Glycan Composition Profiling for Quality Testing of Biotherapeutics. BioPharm International. 2015;28(12):59-64.

[10] Zhao S, Walsh I, Abrahams JL, Royle L, Nguyen-Khuong T, Spencer D, et al. GlycoStore: a database of retention properties for glycan analysis. Bioinformatics. 2018;34(18):3231-2.

[11] Rudd PM, Shi X, Taron CH, Walsh I. Recent Advances in the Use of Exoglycosidases to Improve Structural Profiling of N-glycans from Biologic Drugs. BioPharm International 2018;31 (10):16-23.

[12] Walsh I, Nguyen-Khuong T, Wongtrakul-Kish K, Tay SJ, Chew D, José T, et al. GlycanAnalyzer: software for automated interpretation of N-glycan profiles after exoglycosidase digestions. Bioinformatics. 2019;35(4):688-90.

[13] Kozak RP, Royle L, Gardner RA, Fernandes DL, Wuhrer M. Suppression of peeling during the release of O-glycans by hydrazinolysis. Anal Biochem. 2012;423(1):119-28.

[14] Merry AH, Neville DC, Royle L, Matthews B, Harvey DJ, Dwek RA, et al. Recovery of intact 2-aminobenzamide-labeled O-glycans released from glycoproteins by hydrazinolysis. Anal Biochem. 2002;304(1):91-9.

[15] Wilkinson H, Saldova R. Current Methods for the Characterization of O-Glycans. Journal of Proteome Research. 2020;19(10):3890-905.

[16] Rogers RS, Abernathy M, Richardson DD, Rouse JC, Sperry JB, Swann P, et al. A View on the Importance of "Multi-Attribute Method" for Measuring Purity of Biopharmaceuticals and Improving Overall Control Strategy. AAPS J. 2017;20(1):7.

[17] Giansanti P, Tsiatsiani L, Low TY, Heck AJR. Six alternative proteases for mass spectrometry–based proteomics beyond trypsin. Nature Protocols. 2016;11(5):993-1006.

[18] Yang S, Wu WW, Shen R, Sjogren J, Parsons L, Cipollo JF. Optimization of O-GIG for O-Glycopeptide Characterization with Sialic Acid Linkage Determination. Anal Chem. 2020;92(16): 10946-51.

[19] Suttapitugsakul S, Sun F, Wu R. Recent Advances in Glycoproteomic Analysis by Mass Spectrometry. Anal Chem. 2020;92(1):267-91.

[20] Chen CC, Su WC, Huang BY, Chen YJ, Tai HC, Obena RP. Interaction modes and approaches to glycopeptide and glycoprotein enrichment. Analyst. 2014;139(4):688-704.

[21] Parker BL, Thaysen-Andersen M, Solis N, Scott NE, Larsen MR, Graham ME, et al. Site-Specific Glycan-Peptide Analysis for Determination of N-Glycoproteome Heterogeneity. Journal of Proteome Research. 2013; 12(12):5791-800.

[22] Xu Y, Wu Z, Zhang L, Lu H, Yang P, Webley PA, et al. Highly specific enrichment of glycopeptides using boronic acid-functionalized mesoporous silica. Anal Chem. 2009;81(1):503-8.

[23] Larsen MR, Jensen SS, Jakobsen LA, Heegaard NH. Exploring the sialiome using titanium dioxide chromatography and mass spectrometry. Mol Cell Proteomics. 2007;6(10):1778-87.

[24] Zhang H, Li XJ, Martin DB, Aebersold R. Identification and quantification of N-linked glycoproteins using hydrazide chemistry, stable isotope labeling and mass spectrometry. Nat Biotechnol. 2003;21(6):660-6.

[25] Nilsson J, Rüetschi U, Halim A, Hesse C, Carlsohn E, Brinkmalm G, et al. Enrichment of glycopeptides for glycan structure and attachment site identification. Nature Methods. 2009; 6(11):809-11.

[26] Yang W, Ao M, Hu Y, Li QK, Zhang H. Mapping the O-glycoproteome using site-specific extraction of O-linked glycopeptides (EXoO). Mol Syst Biol. 2018;14(11):e8486.

[27] Ruiz-May E, Catalá C, Rose JK. N-glycoprotein enrichment by lectin affinity chromatography. Methods Mol Biol. 2014;1072:633-43.

[28] Ma J, Hart GW. O-GlcNAc profiling: from proteins to proteomes. Clin Proteomics. 2014;11(1):8.

[29] Sankaranarayanan R, Sekar K, Banerjee R, Sharma V, Surolia A, Vijayan M. A novel mode of carbohydrate recognition in jacalin, a Moraceae plant lectin with a beta-prism fold. Nat Struct Biol. 1996;3(7):596-603.

[30] Wang K, Peng ED, Huang AS, Xia D, Vermont SJ, Lentini G, et al. Identification of Novel O-Linked Glycosylated Toxoplasma Proteins by *Vicia villosa* Lectin Chromatography. PloS one. 2016;11(3):e0150561-e.

[31] Fanayan S, Hincapie M, Hancock WS. Using lectins to harvest the plasma/serum glycoproteome. Electrophoresis. 2012;33(12):1746-54.

[32] Zielinska DF, Gnad F, Wiśniewski JR, Mann M. Precision mapping of an in vivo N-glycoproteome reveals rigid topological and sequence constraints. Cell. 2010;141(5):897-907.

[33] Mizushima T, Hirao T, Yoshida Y, Lee SJ, Chiba T, Iwai K, et al. Structural basis of sugar-recognizing ubiquitin ligase. Nat Struct Mol Biol. 2004;11(4): 365-70.

[34] Mizushima T, Yoshida Y, Kumanomidou T, Hasegawa Y, Suzuki A, Yamane T, et al. Structural basis for the selection of glycosylated substrates by SCF(Fbs1) ubiquitin ligase. Proc Natl Acad Sci U S A. 2007;104(14):5777-81.

[35] Yoshida Y, Mizushima T, Tanaka K. Sugar-Recognizing Ubiquitin Ligases:

Action Mechanisms and Physiology. Front Physiol. 2019;10:104.

[36] Hagihara S, Totani K, Matsuo I, Ito Y. Thermodynamic Analysis of Interactions between N-Linked Sugar Chains and F-Box Protein Fbs1. Journal of Medicinal Chemistry. 2005;48(9):3126-9.

[37] Chen M, Shi X, Duke RM, Ruse CI, Dai N, Taron CH, et al. An engineered high affinity Fbs1 carbohydrate binding protein for selective capture of N-glycans and N-glycopeptides. Nat Commun. 2017;8:15487.

[38] Juillerat A, Gronemeyer T, Keppler A, Gendreizig S, Pick H, Vogel H, et al. Directed evolution of O6-alkylguanine-DNA alkyltransferase for efficient labeling of fusion proteins with small molecules in vivo. Chem Biol. 2003; 10(4):313-7.

[39] Keppler A, Gendreizig S, Gronemeyer T, Pick H, Vogel H, Johnsson K. A general method for the covalent labeling of fusion proteins with small molecules in vivo. Nat Biotechnol. 2003;21(1):86-9.

[40] Ganatra MB, Potapov V, Vainauskas S, Francis AZ, McClung CM, Ruse CI, et al. A bi-specific lectin from the mushroom Boletopsis grisea and its application in glycoanalytical workflows. Sci Rep. 2021;11(1):160.

[41] Carrizo ME, Capaldi S, Perduca M, Irazoqui FJ, Nores GA, Monaco HL. The Antineoplastic Lectin of the Common Edible Mushroom (Agaricus bisporus) Has Two Binding Sites, Each Specific for a Different Configuration at a Single Epimeric Hydroxyl*. Journal of Biological Chemistry. 2005;280(11):10614-23.

[42] Leonidas DD, Swamy BM, Hatzopoulos GN, Gonchigar SJ, Chachadi VB, Inamdar SR, et al. Structural Basis for the Carbohydrate Recognition of the Sclerotium rolfsii Lectin. Journal of Molecular Biology. 2007;368(4):1145-61.

[43] Riley NM, Hebert AS, Westphall MS, Coon JJ. Capturing site-specific heterogeneity with large-scale N-glycoproteome analysis. Nat Commun. 2019;10(1):1311.

[44] Riley NM, Westphall MS, Hebert AS, Coon JJ. Implementation of Activated Ion Electron Transfer Dissociation on a Quadrupole-Orbitrap-Linear Ion Trap Hybrid Mass Spectrometer. Anal Chem. 2017;89(12):6358-66.

[45] Saba J, Dutta S, Hemenway E, Viner R. Increasing the productivity of glycopeptides analysis by using higher-energy collision dissociation-accurate mass-product-dependent electron transfer dissociation. Int J Proteomics. 2012;2012:560391.

[46] Singh C, Zampronio CG, Creese AJ, Cooper HJ. Higher energy collision dissociation (HCD) product ion-triggered electron transfer dissociation (ETD) mass spectrometry for the analysis of N-linked glycoproteins. J Proteome Res. 2012;11(9):4517-25.

[47] Bern M, Kil YJ, Becker C. Byonic: advanced peptide and protein identification software. Curr Protoc Bioinformatics. 2012;Chapter 13: Unit13.20.

[48] Toghi Eshghi S, Shah P, Yang W, Li X, Zhang H. GPQuest: A Spectral Library Matching Algorithm for Site-Specific Assignment of Tandem Mass Spectra to Intact N-glycopeptides. Analytical Chemistry. 2015;87(10):5181-8.

[49] Liu M-Q, Zeng W-F, Fang P, Cao W-Q, Liu C, Yan G-Q, et al. pGlyco 2.0 enables precision N-glycoproteomics with comprehensive quality control and one-step mass spectrometry for intact glycopeptide identification. Nature Communications. 2017;8(1):438.

[50] Lu L, Riley NM, Shortreed MR, Bertozzi CR, Smith LM. O-Pair Search with MetaMorpheus for O-glycopeptide characterization. Nat Methods. 2020;17(11):1133-8.

[51] Zhan T, Rindtorff N, Boutros M. Wnt signaling in cancer. Oncogene. 2017;36(11):1461-73.

[52] Sengupta PK, Bouchie MP, Kukuruzinska MA. N-glycosylation gene DPAGT1 is a target of the Wnt/beta-catenin signaling pathway. J Biol Chem. 2010;285(41):31164-73.

[53] Jamal B, Sengupta PK, Gao ZN, Nita-Lazar M, Amin B, Jalisi S, et al. Aberrant amplification of the crosstalk between canonical Wnt signaling and N-glycosylation gene DPAGT1 promotes oral cancer. Oral Oncol. 2012;48(6):523-9.

Section 4

Glycans

The Structure of Leukocyte Sialic Acid-Containing Membrane Glycoconjugates is a Differential Indicator of the Development of Diabetic Complications

Iryna Brodyak and Natalia Sybirna

Abstract

Glycans, as potential prognostic biomarkers, deserve attention in clinical glycomics for diseases diagnosis. The variety of glycan chains, attached to proteins and lipids, makes it possible to form unique glycoconjugates with a wide range of cellular functions. Under leukocyte-endothelial interaction, not only the availability of glycoconjugates with sialic acids at the terminal position of glycans are informative, but also the type of glycosidic bond by which sialic acids links to subterminal carbohydrates in structure of glycans. The process of sialylation of leukocyte glycoconjugates undergoes considerable changes in type 1 diabetes mellitus. At early stage of disease without diabetic complications, the pathology is accompanied by the increase of α2,6-linked sialic acids. The quantity of sialic acid-containing glycoconjugates on leukocytes surface increases in condition of disease duration up to five years. However, the quantity of sialic acids linked by α2,6-glycosidic bonds decreases in patients with the disease duration over ten years. Therefore, sialoglycans as marker molecules determine the leukocyte function in patients with type 1 diabetes mellitus, depending on the disease duration. Changes in the glycans structure of membrane glycoconjugates of leukocytes allow understanding the mechanism of diabetic complications development.

Keywords: sialic acid, glycans, glycoconjugates, sialyltransferases, sialidases, leukocytes, type 1 diabetes mellitus

1. Introduction

Glycans, as potential biomarkers of health and illness, deserve attention in clinical glycomics for early-stage disease diagnosis [1, 2]. It is not surprisingly that abnormal (aberrant) glycosylation of proteins and lipids have been observed in many diseases, including cancer, cardiovascular disease, immune deficiencies and diabetes [3].

Diabetes mellitus (DM) is one of the most common endocrine diseases that have been identified as one of the priority issues for national health systems around the world. Type 1 DM is characterized by a progressive autoimmune destruction of pancreatic β-cells, leading to insulin deficiency and chronic hyperglycemia [4]. The social significance of this problem is that DM associated with development of numerous concomitant diseases, early disability, and metabolic complications [5]. Insufficient control of glucose level in blood may increased the risk of microvascular (nephropathy, retinopathy, neuropathy) and macrovascular (peripheral artery disease, coronary artery diseases, congestive heart failure, myocardial infarction, stroke) complications. In addition, individuals with DM have increased risk of physical and cognitive disability, depression and cancer [6]. The various complications related to diabetes are determined by changes of blood components. Blood cells (leukocytes, erythrocytes and thrombocytes) are of particular interest because they are directly exposed to high glucose concentrations [7]. Blood cells aggregate strongly to the vessel wall and adhere to each other, which leads to the development of pathological changes in capillary blood flow and microcirculation disorders in diabetes [4–6].

Leukocytes are the major cells of the inflammatory and immune response that defends against different type of infection, consequently these cells are important object for investigation [5]. The clinical and experimental studies of human blood in case of DM and animals with streptozotocin-induced diabetes demonstrate significant violations of the morphofunctional state of leukocytes. The dysfunction of chemotaxis capacity, adhesion and migration, reduction of phagocytic activity and bactericidal ability of leukocytes correlate with the level of hyperglycemia in blood [8, 9].

The impairment in the functions of immunocompetent cells leads to a decrease in immune defense and the development of chronic infectious/inflammatory processes in organism of people with type 1 diabetes [10]. Chronic inflammation is the main cause of progression of diabetic complications which leads to dysfunction of the extremities, retina, kidneys, nerves, heart and blood vessels. According to statistics, most of patients die from angiopathic complications of diabetes. Screening of complications provides with possibility to reduce the risk for their development and progression [11]. Therefore, expansion of diagnostic methods for characterization of changes in the morphofunctional state of leukocytes and the search for preventive remedies that would ameliorate the clinical condition of patients is a relevant problem today.

2. Importance of membrane glycoconjugates in providing the functional activity of leukocytes

Experimental studies have shown that cells of the immune system are exposed to the direct and indirect effects of high blood glucose concentrations in patients with diabetes [4, 5]. Glucose metabolism pathways are activated under conditions of hyperglycemia include the autooxidation of glucose, caused glycation of long-lived proteins; the hexosamine pathway, which leads to the glycosylation of hydroxyl-containing amino acid residues; sorbitol metabolism, accompanied by the formation of free radical; and oxidative phosphorylation leading to mitochondria electron transport chain intensification and the generation of superoxide-anion radicals [12, 13]. Glucose autooxidation and glycation of proteins and lipids leads to an accumulation of advanced oxidation protein products and advanced glycation end products (AGEs), which are difficult to eliminate from the blood and remain in circulation [14]. They are also the source of reactive oxygen species (ROS) since they imitate metal containing oxidation systems. Excessive formation of ROS and reactive nitrogen species (RNS) leads to the development of oxidative-nitrative stress [15]. These changes create a favorable background for the formation of micro- and macrovascular diabetic complications [15, 16].

In condition of hyperglycemia, leukocytes are preactivated by ROS and RNS, angiotensin II, and AGEs. The interaction of AGEs with their receptors, RAGE, causes intracellular signal transduction, which leads to changes in gene expression, overproduction of free radicals, the release of pro-inflammatory molecules (tumor necrosis factor α (TNFα), interleukin 1β (IL-1β), IL-2, IL-6 etc.), and changes in the activity of intracellular enzymes [17].

Glycosyltransferases and glycosidases, which involved in the synthesis of glycans of glycoconjugates, pay much interest among cellular enzymes [18]. In general, mammalian glycans are the product of several types of glycohydrolases and dozens of glycosyltransferases, which act sequentially in the process of oligosaccharide chains synthesis. Each of the glycosyltransferases uses one type of sugar substrate and forms a specific bond between one monosaccharide and a glycan precursor. Thus, the set of glycosyltransferases in the cell determines what type of glycans, among the large number of possible structures, will be formed [19].

The variety of monosaccharides is very large, but most often the carbohydrate components of glycoconjugates of eukaryotic cells include glucose (Glc), N-acetylglucose (GlcNAc), galactose (Gal), N-acetylgalactose (GalNAc), mannose (Man), N-acetylneuraminic acid (Neu5Ac), also known as sialic acid (Sia). Different monosaccharides, combined in a specific sequence by glycosyltransferases, form a glycan, which at one end attaches to a protein or lipid molecule. The formed glycoconjugates are the main macromolecular constituents of biomembranes [19, 20].

The diversity of glycans attached to proteins and lipids makes it possible to form unique glycoconjugates with a wide range of cellular functions. Glycoconjugates play an important role in various biological processes, in particular, glucose homeostasis, protein quality control, cellular differentiation, adhesion, intercellular signaling and inflammation. It is known that carbohydrate residues increase the solubility of glycoproteins, protect against proteolysis, influence on their folding, intracellular transport and secretion. Glycoconjugates are components of the glycocalyx, providing specific interactions with ligands, intercellular contacts and communication. Glycans of glycoconjugate are involved in the formation of the immune response, blood clotting and provide the individuality of organisms and their plasticity [20].

The immune system is highly controlled and fine-tuned by glycosylation, through the addition of a variety of glycans to virtually all adhesion molecules and receptors of leukocytes. Glycoconjugates are implicated in fundamental cellular and molecular processes. Glycans perform function of molecular recognition that regulates both stimulatory and inhibitory immune pathways [21]. The presence of modified carbohydrate determinants in the glycan structure modifies the biological activity of the entire glycoconjugate. The interaction of specific ligand with its modified receptor leads to violations at the level of transmembrane and intracellular signaling [22]. In according to the importance of glycans in the immune system, scientific researches emphasize the essential contributions of glycosylation in the regulation of innate and adaptive immune responses [21]. Therefore, today the scientists (biochemists, molecular biologists, immunologists, pathologists and pharmacologists) are making the great efforts to explore the interrelations of carbohydrate determinants with their glycobiology. Establishing changes in the glycans structure of membrane glycoconjugates of immunocompetent cells makes it possible to understand the mechanism of pathological changes in condition of diabetes and diabetic complications.

Thus, changes in intracellular metabolism, intensification of glycation processes and the development of oxidative-nitrative stress in blood cells under conditions of prolonged hyperglycemia are the main factors that induce pathological changes in the structure of their components and affect their functional state [12, 20].

3. Sialoglycoconjugates of leukocytes as the main regulators of molecular and cellular interactions

Glycoconjugates of leukocytes contain sialic acid as the terminal sugar and play important roles in many physiological processes. Sialic acid normally exists in the periphery of non-reducing end of the oligosaccharide chains of many glycoproteins and glycolipids. They are involved in carbohydrate-protein interactions during cell recognition, in cell–cell interactions involving functional receptors, in the binding of pathogens such as viruses, bacteria or parasites [19, 23]. Sialic acids are also implicated in the processes of activation, differentiation, survival and apoptosis of leukocytes. Sialoglycoconjugates affects cellular adhesiveness, antigenicity, action of some hormones, catalytic properties of enzymes, modulating the affinity of cell surface receptors and transmembrane signaling [19, 20]. It is obvious that sialic acids are important molecular determinants of many immune processes. To implement these functions, organisms have a range of proteins (sialospecific lectins) that recognize surface-exposed sialic acids in glycoconjugates [19].

The variety of functions indicates the importance of sialic acid in cell biology. The biology of sialic acids should be considered from the point of view of their dual function. On the one hand, sialic acid acts as biological mask agent by masking recognition sites such as receptor molecules of cell membranes. On the other hand, sialic acid plays a role as recognizable cell patterns. Sialic acids as ligands are recognized by lectins, antibodies, hormones or as receptors recognize extracellular markers in the molecular processes of cell interactions [23–25]. Activation of cells can lead to the opening of ligand-binding sites with a subsequent increase in binding affinity, lowering the cellular activation threshold, or removal of inhibitory signals [26, 27].

Sialic acids can participate directly or indirectly in multiple cellular events and overall immune response [28]. Sialic acids contribute to cells being "self" and, thus, weakens immunoreactivity. That is why they are not recognized by immune system cells or macrophage lectins. The loss of these masking monosaccharides makes the cell "foreign", activating the body's immunoreactive response. Therefore, sialic acids can be considered components of innate immune protection [29]. These acids have recently been recognized as being involved in most important phenomena of molecular and cellular interactions in immune regulation [30, 31]. In this respect, sialic acids have been associated with inflammatory diseases, malignancies, cardiovascular disease and diabetes [32].

Sialic acids are group of monosaccharides with high structural diversity, which are chemically derived from nine carbon acidic sugars – neuraminic acids. The most abundant member of the family carries an acetyl moiety linked to the amino group of fifth carbon (C5) giving the Neu5Ac. A feature of its structure is the presence of a carboxyl group near C1, which determines the negative charge of the molecule at physiological pH and characterizes it as a strong organic acid (pK 2.2). More than fifty derivatives of neuraminic acids have been found in nature. The most common sialic acid derivatives found in mammals are Neu5Ac and N-glycolylneuraminic acid (Neu5Gc), whereas in humans Neu5Ac is the dominant sialic acid [19, 23, 33].

Due to their negative charge at physiological pH and hydrophilic property sialic acids stabilize conformation of molecules, can impact protein oligomerization, the interactions of proteins with other proteins and the extracellular matrix. Sialic acids as an essential compound of all cell membranes play an important role in maintaining the structure, permeability and integrity of the cell membrane [28, 34]. Not surprisingly, sialic acid exponation is dynamic, changes during development and is altered in numerous diseases [35]. Changes in sialylation are associated with oxidative stress induced by several disorders including diabetes. It has been proven that level of sialic acid increased in plasma in condition of inflammatory processes and DM [36]. The relation between sialic acid and diabetes duration most likely follows from the association sialic acid with microvascular complications, which are well established to be related to glycemic control [37]. Therefore, sialic acid concentrations in the blood may be a useful marker of the development of diabetic complications, but there have been no many studies examining the link between sialoglycoconjugates and complications in type 1 DM.

The structural diversity of sialoglycoconjugates is due not only to the diversity of derivatives of sialic acids in their composition, but also depends on the type of glycosidic linkages (2,3, 2,6, 2,8, and 2,9) with subterminal sugars. The sialylation of oligosaccharide chains of glycoconjugates is carried out with the participation of the family of enzymes sialyltransferases (STs). About 20 STs have been characterized [19]. STs are divided in four main subfamilies, namely the ST3Gal, ST6Gal, ST6GalNAc and ST8Sia, depending on the glycosidic linkage formed and the monosaccharide acceptor recognized [35, 38]. ST3Gal, ST6Gal, ST6GalNAc and ST8Sia link Neu5Ac via its C2 to the C3, C6 positions of other carbohydrates or the C8, C9 positions of another sialic acids, generating $\alpha2,3$-, $\alpha2,6$-, $\alpha2,8$, or $\alpha2,9$-linked sialic acids, respectively [19, 39]. Sialyltransferase-mediated addition of sialic acid on glycans usually stops their further growth and modifies charge, steric hindrance, conformation and flexibility, underlying the importance of STs in shaping the structures and functions of sialoglycans [35, 40].

In the structure of leukocytes' glycans sialic acids are frequently the terminal residues of glycans and are mostly attached either by a 2,3- or 2,6-glycosidic bond to Gal or GalNAc of oligosaccharide chains [19]. The ST6Gal and ST6GalNAc, which are present in leukocytes, catalyze the transfer of Neu5Ac from CMP-Sia (cytidine-5′-monophospho-N-acetylneuraminic acid) to the C6 hydroxyl group of a terminal Gal or GalNAc residues, respectively, with the formation of $\alpha2,6$-linkaged sialic acids in the oligosaccharide chains of glycans [19, 20, 38]. The ST3Gal comprises family with six members (ST3Gal I–VI). The expression of ST6Gal-I is tissue specific and regulated by multiple transcriptional promoters [41, 42]. An inducible and liver-specific promoter drive high ST6Gal-I expression during inflammation with increase in secreted ST6Gal-I in blood [43]. Activated platelets release the CMP-Sia that serves as the donor for circulating ST6Gal-I, allowing for the remodeling of the glycans of hematopoietic stem cells and multipotent progenitors (HSC/MPPs) [44]. Thus, inducible promoter is important for regulation of hematopoiesis [45]. The ST3Gal-V add a sialic acid to terminal Gal residues with the formation of $\alpha2,3$-glycosidic linkage, while ST3Gal-IV sialylates Galβ1,3GalNAc terminated structures in glycoconjugates and Galβ1,4(3)GlcNAc structures found on N- and O-glycans [46]. The ST3Gal-IV and ST3Gal-VI involved in the synthesis of the sialyl Lewis[X] (sLe[X]) determinant on leukocyte E-, L- and P-selectin ligands [19, 46]. Leukocytes express a number of different selectin ligands, including E-selectin ligand-1, P-selectin glycoprotein ligand-1, CD43, CD44, β2-integrins, ets. [35, 47].

Glycosylation of cell-surface structures of leukocytes is important in the accomplishment of the immune function by these cells in organism. The membrane structures of leukocytes are decisive in the processes of extravasation, the migration of leukocytes from blood vessels into the extracellular space. In order to penetrate the vascular wall, leukocytes initially interact with the endothelium, roll over its surface, undergo dense adhesion, dissolve, and finally move through or between endothelial cells of the blood vessel [48–50]. Leukocyte chemotaxis depends on the surface sLe^X and E-selectin of vascular endothelial cells. E-, L- and P-selectins are exposed by endothelial cells, leukocytes and platelets, respectively. Selectins are carbohydrate-binding proteins that recognize the sLe^X structure (Neu5Acα2,3Galβ1,4(Fucα1,3)GlcNAcβ-R), capping N- and O-glycans as specific ligands [51–53]. These sialic acid-containing moieties are required for leukocyte binding to selectins on endothelial cells and their rolling [54, 55]. Combinatorial knockout of ST3Gal-IV and ST3Gal-VI that are the involved in sLe^X synthesis leds to a decrease in neutrophil binding to E- and P-selectins, selectin-dependent rolling, and lymphocyte homing [46]. The selectin profile of cells can change under the influence of cytokines in case of development of inflammatory process, infection or under the influence of ROS. In tissue inflammation, cytokines stimulated endo-thelial cell production of E-selectin, which could recognize sLe^X on the leukocyte surface and bind it, promoting leukocyte adhesion to the vascular endothelium and, subsequently, to the inflammatory tissue or locations of injury [32, 56]. Therefore, the recognition of all types of selectins is mediated with sialic acid residues [19].

In the catabolism of sialoglycans of glycoconjugate involved extracellular and intracellular sialidases, a glycoside hydrolase, that specifically hydrolyze release α-linked sialic acid residues through hydrolysis of the glycosidic bond between the acidic sugar(s) and the internal acceptor. Four different sialidases (also termed as neuraminidases – NEUs) in mammalian cells, NEU1, NEU2, NEU3 and NEU4, have been described [57]. These NEUs exhibit differences in cellular localization, substrate specificities, physiological functions and expression patterns in different tissues and physio/pathological conditions [35, 57, 58]. The NEU1 is found in the lysosome and on the cell surface and is the most highly expressed of this sialidase family [35]. The level of NEU2 is extremely low and the content of NEU3 and NEU4 are about 10% of NEU1 in tissue separately [57]. The lysosomal sialidase NEU1 initiates the degradation of sialoglycoconjugates [59]. The NEU1 is capable of hydrolyzing a wide range of glycoproteins, oligosaccharides and ganglioside near neutral pH. It exclusively acts on glycoproteins and preferentially cleaves α2,3-linkages over α2,6- or α2,8-linkages [19, 35]. In addition, NEU1 may have extralyso-somal localization and focus on the periphery of activated lymphocytes. The NEU1 controls several aspects of the immune response by the desialylation of molecules, such as Toll-like receptor 4 and adhesion molecules involved in the recruitment of leukocytes to inflammatory sites [35, 57]. Desialylation of sialyl α2,3-linked Gal residues of Toll-like receptor 4 is essential for receptor activation and cellular signaling [60]. The cytosolic sialidase (NEU2) can hydrolyze sialic acids from glycoproteins and gangliosides [61]. The plasma membrane-associated sialidase (NEU3) is a key enzyme for ganglioside hydrolysis [57]. The NEU4 is localizing in the lysosomal lumen or bound to the outer mitochondrial membranes via pro-tein–protein interactions or the ER membrane-associated. Its exhibits the highest activity with gangliosides as well as sLe^X and sLe^a antigens [35, 57, 58]. Sialic acid is actively exfoliated from the cell surface by extracellular sialidases during leukocyte activation. This process plays an important regulatory role in cell activation and differentiation [62].

Metabolism of sialic acids includes the cooperation of enzymes that catalyze the biosynthesis, activation, transfer of sialic acids to glycoconjugates, as well as the removal and degradation of sialic acids [63]. The aberrant expression of STs and NEUs accelerates and sustains sialylation status on glycoconjugates [64]. Therefore, knowledge in this field of glycobiology allows to predict biological events in case of increase or decrease in the amount of sialoglycoconjugates on the cell surface or under conditions of modification or structural changes of these acids in certain types of cells. Thus, STs, NUEs and sialic acids itself represent important therapeutic targets for medicinal chemistry and biopharmaceutical industry [65, 66].

4. Lectins as diagnostic molecular probes for determining the glycosylation profile and structural changes of glycans

Glycocode information is read in living organisms with the help of specific compounds – lectins. Lectins are sugar-binding proteins that can specifically recognize glycans of glycoconjugates without disrupting the structure of the recognizable carbohydrate-containing ligands.

Since surface glycoconjugates have a unique structure for each cell type, they can be identified, quantified and characterised structural changes in glycans using specific lectins. Nowadays, lectins, their properties, the importance of these proteins in the life of organisms and their applying in experimental biology and medicine are the subject of research in the world's science laboratories. Lectins excluded from living objects are valuable biochemical reagents that are used in experimental cytochemistry, in the diagnosis of some diseases, and in biotechnology for isolating certain carbohydrate-containing molecules [20, 67].

Interactions of sialic acids with lectins play a leading role in many physiological and pathological processes. Therefore, sialospecific lectins are used to recognize sialic acids with specific linkages to subterminal sugars. Wheat germ lectin (WGA) specifically binds to β,DGlcNAc and Neu5Ac. The *Maackia amurensis* lectin (MAA) and *Sambucus nigra* lectin (SNA) are commonly used to recognize the α2,3-linked (Neu5Acα2,3Gal) and α2,6-linked (Neu5Acα2,6Gal/GalNAc) sialic acid residues, respectively [20, 68, 69]. Sialospecific lectins apply in lectin microarray [70], histochemistry [71], in lectin blot [72, 73], fluorescent image and flow cytometry [74] (**Figure 1**). At the same time, the combination of lectins with monoclonal antibodies can be used to obtain complete information on the antigenic repertoire of cells both in normal and in case of pathologies [73].

Blood leukocytes are similar in structural organization, and, at the same time, they differ significantly in biochemical structure. It is very important to understand the morphofunctional state of the cell is to be able to detect these differences. Numerous methods are used for this purpose [70]. Aggregatometry is one of the assays used to evaluate the functional properties of platelets, leukocytes and erythrocytes in the dynamics, monitor antiplatelet therapy, study the mechanisms of aggregation. The aggregation capacity of cells is assessed by such parameters as the degree, rate and time of aggregation [20].

The substances of protein (lectins, proteolytic enzymes, chemoactive peptides); lipid (metabolites of arachidonic acid, liposomes); carbohydrate (heparin, dextransulfates) or other nature (phorbol esters, amphotericin B, ADP, organic dyes – alcyanine blue, ruthenium red) can be inducers of aggregation. Lectins used in aggregatometry are divided into lectins-mitogens (ConA, PHA) and polyvalent lectins (WGA, SNA). Polyvalent lectins have two or more binding centers of carbohydrate determinants (carbohydrate-recognition domains) on the cell surface. The aggregation of cells by such lectins is due to the formation of intercellular molecular bridges. The ability of each subunit of lectin to bind sugars individually leads to the formation of a cross-linked structure of the aggregate. The efficiency of lectin-induced aggregation is determined by the processes of clustering of lectin receptors on the cell surface [20, 75].

The aggregation capacity of leukocytes is studied to model their pre-migratory state before leaving the bloodstream, i.e. before diapedesis, or to analyze phagocytic activity. It is consider that phagocytosis involving lectin-carbohydrate interactions

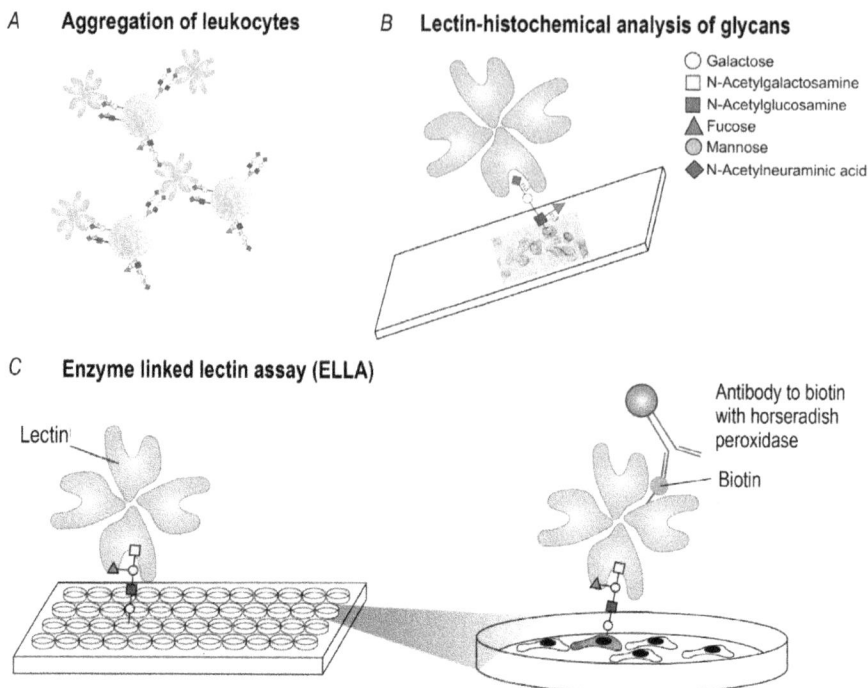

Figure 1.
Examples of uses of lectins in glycobiology. Many plant and animal lectins are multivalent. In particular, the lectin is shown with four carbohydrate binding domains. (A) Lectins bind of surface glycoconjugates of leukocytes, causes cell aggregation. (B) Histochemical analysis of surface glycans. (C) Enzyme linked lectin assay: biotinylated lectins bind to glycoconjugates on the surface of cells immobilized to the bottom of the well of a flat-bottomed plate; bound lectins are detected by antibodies to biotin with horseradish peroxidase.

is one of the oldest evolutionary forms of this process [76]. Phagocytosis during evolution was significantly displaced by antigen–antibody interactions, but did not lose importance in the formation of a nonspecific immune response [77].

5. Changes of carbohydrate determinants of glycoconjugate mediate the functional state of leukocyte in type 1 DM

Leukocytes are markers of the immune homeostasis and receive signals from the microenvironment through the glycans of receptors. Lectins of certain carbohydrate specificity are ligands that selectively activate chemokine receptors. The response of cells to lectins *in vitro* makes it possible to analyze the chemical structure of the carbohydrate determinants of glycoconjugates on the membrane of leukocytes [66, 78].

Lectins WGA, SNA and MAA, which specific to sialic acids, are used to determination sialylated glycoconjugates and differentiation various types of sialic acid links with subterminal carbohydrates of glycans (SNA recognizes α2,6-links, while MAA identifies α2,3-links, **Figure 2**) [75, 79].

Alteration of the amount of sialic acids on the surface of leukocytes is an additional level of regulation of cells affinity to signaling molecules (cytokines, hormones), pathogenic microorganisms and determines the nature of cell–cell interactions [20].

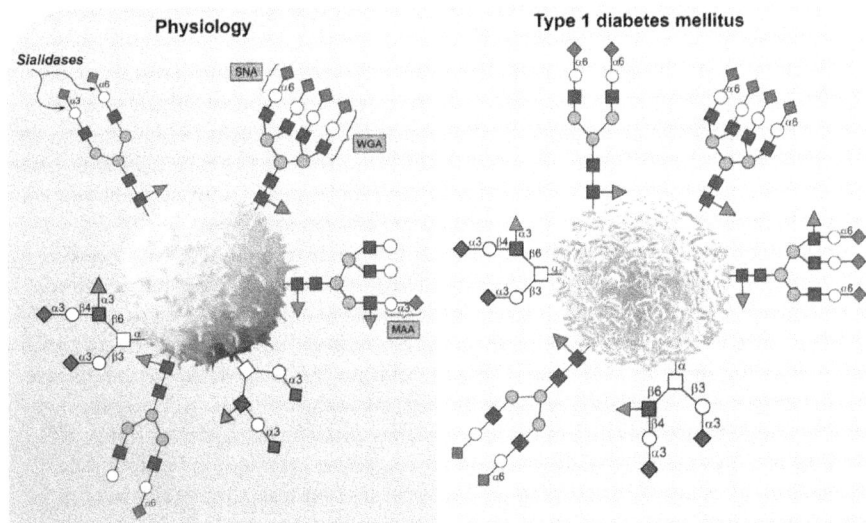

Figure 2.
The structure of leukocyte sialic acid-containing membrane glycans in physiological state of cells and in type 1 diabetes. Sialic acids, depending on the type of glycosidic bond in the structure of the glycan, are recognized by WGA, SNA and MAA lectins.

The most significant changes of increasing lectin-induced aggregation of leukocytes in type 1 diabetes have been observed using lectin WGA. An increase in the degree and rate of WGA-induced aggregation of neutrophils in diabetes is a sign of increased of N-acetyl-β,D-glucosamine-containing and sialic acid-containing glycoconjugates on surface of leukocytes [80]. This indicates that synthesis of hybrid types of N-glycans is occured by activated N-acetylglucosaminyltransferase-III (GnT-III) and incomplete glycosylation of proteins and lipids [20]. As a result, glycoconjugates with terminal β,D-GlcNAc residues are exhibited on the leukocyte surface and determined high rates of WGA-induced aggregation [81].

The structural characterization of neutrophils glycoconjugates showed that cell surface N-glycans are highly sialylated, and many of their "antenna" play an important role in selectin-mediated neutrophil circulation [82]. Glycome of neutrophils is consisted mainly of complex bi- tri- and tetra-antennary N-glycans (**Figure 2**). Their antennae are predominantly terminated with Neu5Ac and Le^X (Galβ1,4(Fucα1,3)GlcNAc) epitopes [83]. The ST3Gal-IV knockout results in significant reduction in the synthesis of sLe^X structures in neutrophils. These cells show significant impairment in rolling and adhesion to the endothelial cells [84]. All these structural changes in the carbohydrate chains of glycoconjugates of leukocytes induce disturbances of molecular signals perception from the microenvironment, affecting interaction of leukocytes with other circulating blood cells and vascular endothelium in condition of diabetes [7, 20, 85].

Sialic acids can mask, i.e. change the structure of carbohydrate components of various specific receptors on the cell surface [19]. There is the receptor to N-formyl-methionyl-leucil-phenylalanine, C5a component of the complement system, IL-8, the receptor of granulocyte-monocyte colony-stimulating factor and the cell receptor 3 (Mac-1) among WGA-binding glycoproteins. The interaction of neutrophils with the intercellular adhesion molecule 1 (ICAM-1, CD54), which is involved in the adhesion of leukocytes to the vascular endothelium occurs via the Mac-1 receptor. On the other hand, WGA-specific receptors are involved in the stimulation of respiratory burst in neutrophils by activating NADPH oxidase and followed formation of ROS [86–88].

The content of GlcNAc and Neu5Ac residues in glycans of glycoconjugates of the plasma membrane of neutrophils increases in type 1 diabetes [72]. It may be one of the main causes of nonspecific damage of tissues and cells, which are close to stimulated neutrophils. Under such conditions, neutrophils produce ROS and cause erythrocytes, platelets, fibroblasts and endotheliocytes death, inactivate enzymes, lead to changes in the structure of proteins and lipid peroxidation [6, 20].

Interaction of glycoconjugates of polymorphonuclear leukocytes with lectin SNA changes significantly under DM [80]. The level, velocity and time required for the maximum neutrophilic granulocyte aggregation in patients with type 1 DM duration of up to 5 years have been different from these indicators in patients with diabetes lasting more than 10 years. In particular, in the early stages of the disease, the degree of neutrophils aggregation, as well as the rate of SNA-induced aggregation have been four times higher than in patients with the disease over ten years [20, 80]. It is assumed that with the disease progresses, changes in leukocytes are associated with neutrophil subactivation processes that lead to the release of granule contents into the extracellular space, especially intravascularly. Degranulation leads to the lowering of cells aggregation [88, 89]. It is known that elevated glucose levels inside the cell have an inhibitory effect on a number of enzymes that are involved in the biosynthesis of the oligosaccharide chain of glycans. One of such enzymes is STs, which catalyze the attachment of sialic acid to the terminal sugar in glycan structure [19, 39]. Hyperglycemia is probably one of the factors that mediates the glycan profile violation of leukocytes in diabetes.

Decreased aggregation of neutrophils in patients with DM under the addition of MAA lectin indicates the presence sialic acids in the structure of glycoconjugates of neutrophil membranes in a small amount. These α2,3-linked sialic acids affect both the dynamic and kinetic parameters of the neutrophil aggregation process [72, 90]. The decrease in sialic acid content in the cellular glycocalyx is most often due to the enhanced desialylation of the membrane glycoconjugate. It is worth noting that sialic acids which are linked to subterminal sugars of the glycoconjugates oligosaccharide chains by the α2,3-glycoside bond are much more likely to undergo hydrolytic cleavage by sialidases than α2,6-linked residues of these sugars [90]. Cleavage of sialylated oligosaccharide fragments from glycoconjugates or exfoliation of the whole molecules of sialoglycoconjugates can be another reason of loss of sialic acids from the cell surface. However, there is often a combination of all these factors [20, 81]. Decreased α2,3-linked sialic acid on the surface of leukocytes leads to impaired perception of signals from the extracellular space, interaction with other cells, as well as numerous bacteria, protozoa and viruses. Desialylation of surface glycoconjugates of polymorphonuclear leukocytes leads to increasing of their adhesive properties, which promotes the migration of neutrophils through the vascular endothelium [91].

The interaction of glycoconjugates of mononuclear leukocytes with lectin MAA, which reacts with Neu5Acα2,3 Gal/GalNAc terminal endings glycan, have been markedly inhibited under diabetes [90, 92]. It has been found that the decrease in sialic acid content usually occurs due to increased activity of endogenous sialidases in activated T cells and monocytes [46]. This leads to increased production of cytokines by lymphocytes and interaction of monocytes with hyaluronic acid – a component of extracellular matrix [93, 94]. The NEU1 and NEU3 are expressed in monocytes in the process of their differentiation into macrophages. Desialylation of glycans on the surface of monocytes by exogenous NEU resulted in activation of ERK1/2 and p38 MARK signaling pathways and increased production of IL-6, IL-1β, MIP-1α and MIP-1β [94, 95]. Pro-inflammatory cytokines cause endothelial dysfunction by increasing capillary permeability, inducing prothrombotic properties, promoting leukocyte recruitment by synthesis of adhesion molecules and chemoattractants, and play a role in macroangiopathy by promoting dyslipidemia. Thus, it is unlikely that the increased circulation of sialic acid is the result of desialylation of glycoconjugates. However, there is evidence that sialic acid is reduced in endothelium and erythrocytes in diabetes, which may be important in the pathophysiology of vascular disease [37].

Due to fact that terminal α2,3-linked sialic acids are included, in particular, in the structure of the CD45 receptor, which mediated an increasing of T cell proliferation [96], the decrease content of sialic acids in type 1 diabetes indicates a violation of this function in immunocompetent blood cells. Studies showed that the sialylation of T cell CD45 by ST6Gal-I blocks galectin-1 clustering of CD45 and resulting cell death [97]. The α2,6-sialylation of FasR blocks binding of Fas-associated adaptor molecule to the FasR death domain, thus inhibiting the formation of the death-inducing signaling complex [98].

Lectins SNA and MAA interact with CD45⁺ leukocytes [96]. CD45 is a transmembrane glycoprotein found on T, B, NK cells, granulocytes, and monocytes. It has a cytoplasmic tail with cytosolic phosphotyrosine phosphatase activity. CD45 is the antagonist of tyrosine kinase of insulin receptor, whereas it can show high activity towards membrane-bound molecules (receptors of insulin and epidermal growth factor) [96, 99]. The increased content of sialoglycans in CD45 may cause masking of insulin receptors on organs and tissues, preventing the effect of minimal amounts of the hormone, which can still be produced in type 1 diabetes. This effect may disimprove complications during the development of the disease [96].

The α2,6-sialylation of leukocyte glycoconjugates undergoes certain changes in type 1 DM (**Figure 2**) [100]. Therefore, the quantity of sialic acids linked by α2,6-glycosidic bonds correlate with the disease duration. The content of sialoglycoconjugants on leukocytes surfaces increases for patients with the disease up to five years, while it decreases for patients with the disease duration over ten years. The pathology is accompanied by an increase of linkage places for SNA, which indicates the replacement of α2,3-linked sialic acids by α2,6-linked acids. It is likely to as a result of quantitative changes in the cells or in the enzyme activity of ST6Gal and/or ST6GalNAc [45]. The activity of α2,6 sialyltransferase decreases during the biosynthesis of O-glycans of T lymphocyte in the process of their activation. Thus, an increase in the content of α2,6-linked sialic acids of leukocyte cell surfaces along with a decline in the number of α2,3-linked sialic acids may indicate an increased sensibilization towards B lymphocyte stimulation and the inhibition of T lymphocyte activity under type 1 DM [20, 58].

6. Conclusions

Under cell–cell interaction, not only the presence of certain glycoconjugate, but also the type of linkage of sialic acids to the oligosaccharide chaine is informative. Against the general increase in the number of sialic acid-containing glycoconjugates on leukocytes surface under type 1 DM, there were small quantities of sialic acids linked by α2,3-glycosidic bond to subterminal carbohydrates in structure of glycans. Whereas, the quantity of sialic acid linked by α2,6-glycosidic bonds in the structure of sialoglycans correlated with the duration of diabetes. Such peculiarities of the structure of sialoglycoconjugates of leukocytes may affect both dynamic and kinetic indices of cell aggregation. Leukocytes aggregation affected by lectins may be used as a model of adhesion and migration of these cells. The abnormal redistribution of glycoconjugates on leukocytes membrane under type 1 DM causes changes in their aggregation and adhesion to the vascular endothelium, as well as impairment of the phagocytic function of neutrophils. Thus, the accumulation of leukocyte aggregates in microvessels and violation of disaggregation mechanisms lead to damage of blood vessels. Such changes are etiological preconditions for the development of complications and chronic diseases resulting in deterioration in diabetics' conditions.

Conflict of interest

The authors declare no conflict of interest.

Author details

Iryna Brodyak* and Natalia Sybirna
Department of Biochemistry, Ivan Franko National University of Lviv, Lviv, Ukraine

*Address all correspondence to: iryna.brodyak@lnu.edu.ua

IntechOpen

References

[1] Springer SA, Gagneux P. Glycomics: revealing the dynamic ecology and evolution of sugar molecules. J Proteomics. 2016;135:90-100. DOI: 10.1016/j.jprot.2015.11.022

[2] Svarovsky SA, Joshi L. Cancer glycan biomarkers and their detection – past, present and future. Anal Methods. 2014;6(12):3918-36. DOI: 10.1039/C3AY42243G

[3] Mechref Y, Hu Y, Garcia A, Zhou S, Desantos-Garcia JL, Hussein A. Defining putative glycan cancer biomarkers by MS. Bioanalysis. 2012;4(20):2457-69. DOI: 10.4155/bio.12.246

[4] Goyal SN, Reddy NM, Patil KR, Nakhate KT, Ojha S, Patil CR, et al. Challenges and issues with streptozotocin-induced diabetes – A clinically relevant animal model to understand the diabetes pathogenesis and evaluate therapeutics. Chem Biol Interact. 2016;244:49-63. DOI: 10.1016/j.cbi.2015.11.032

[5] Herold KC, Vignali DAA, Cooke A, Bluestone JA. Type 1 diabetes: translating mechanistic observations into effective clinical outcomes. Nat Rev Immunol. 2013;13(4):243-56. DOI: 10.1038/nri3422

[6] Zhang P, Li T, Wu X, Nice EC, Huang C, Zhang Y. Oxidative stress and diabetes: antioxidative strategies. Front Med. 2020;14(5):583-600. DOI: 10.1007/s11684-019-0729-1

[7] Ma J, Hart GW. Protein O-GlcNAcylation in diabetes and diabetic complications. Expert Rev Proteomics. 2013;10(4):365-80. DOI: 10.1586/14789450.2013.820536

[8] Apostolopoulou M, Menart-Houtermans B, Ruetter R, Nowotny B, Gehrmann U, Markgraf D, et al. Characterization of circulating leukocytes and correlation of leukocyte subsets with metabolic parameters 1 and 5 years after diabetes diagnosis. Acta Diabetol. 2018;55(7):723-31. DOI: 10.1007/s00592-018-1143-x

[9] Menart-Houtermans B, Rütter R, Nowotny B, Rosenbauer J, Koliaki C, Kahl S, et al. Leukocyte profiles differ between type 1 and type 2 diabetes and are associated with metabolic phenotypes: results from the German Diabetes Study (GDS). Dia Care. 2014;37(8):2326-33. DOI: 10.2337/dc14-0316

[10] Marelli-Berg FM, Jangani M. Metabolic regulation of leukocyte motility and migration. J Leukoc Biol. 2018;104(2):285-93. DOI: 10.1002/JLB.1MR1117-472R

[11] Zimmerman RS. Diabetes mellitus: management of microvascular and macrovascular complications. 2016. Available from: https://www.clevelandclinicmeded.com/medicalpubs/diseasemanagement/endocrinology/diabetes-mellitus/

[12] Lozins'ka LM, Semchyshyn HM. Biological aspects of non-enzymatic glycosylation. Ukr Biokhim Zh. 2012;84(5):16-37.

[13] Turk Z. Glycotoxines, carbonyl stress and relevance to diabetes and its complications. Physiol Res. 2010;59(2):147-56. DOI: 10.33549/physiolres.931585

[14] Gradinaru D, Borsa C, Ionescu C, Margina D. Advanced oxidative and glycoxidative protein damage markers in the elderly with type 2 diabetes. J Proteomics. 2013;92:313-22. DOI: 10.1016/j.jprot.2013.03.034

[15] Sifuentes-Franco S, Padilla-Tejeda DE, Carrillo-Ibarra S,

Miranda-Díaz AG. Oxidative stress, apoptosis, and mitochondrial function in diabetic nephropathy. Int J Endocrinol. 2018;2018:1875870. DOI: 10.1155/2018/1875870

[16] Paneni F, Beckman JA, Creager MA, Cosentino F. Diabetes and vascular disease: pathophysiology, clinical consequences, and medical therapy: part I. Eur Heart J. 2013;34(31):2436-43. DOI: 10.1093/eurheartj/eht149

[17] Elbatreek MH, Pachado MP, Cuadrado A, Jandeleit-Dahm K, Schmidt HHHW. Reactive oxygen comes of age: mechanism-based therapy of diabetic end-organ damage. Trends Endocrinol Metab. 2019;30(5):312-27. DOI: 10.1016/j.tem.2019.02.006

[18] Lee C-L, Chiu PCN, Pang P-C, Chu IK, Lee K-F, Koistinen R, et al. Glycosylation failure extends to glycoproteins in gestational diabetes mellitus: evidence from reduced α2-6 sialylation and impaired immunomodulatory activities of pregnancy-related glycodelin-A. Diabetes. 2011;60(3):909-17. DOI: 10.2337/db10-1186

[19] Varki A, Cummings RD, Esko JD, Stanley P, Hart GW, Aebi M, et al. Essentials of Glycobiology. 3rd ed. Cold Spring Harbor (NY): Cold Spring Harbor Laboratory Press; 2015. Available from: http://www.ncbi.nlm.nih.gov/books/NBK310274/

[20] Sybirna NO, editor. Essentials of Glycobiology: monograph. Lviv: Ivan Franko National University of Lviv; 2015. 492 p. (in Ukrainian)

[21] Pereira MS, Alves I, Vicente M, Campar A, Silva MC, Padrão NA, et al. Glycans as key checkpoints of T cell activity and function. Front Immunol. 2018;9:2754. DOI: 10.3389/fimmu.2018.02754

[22] Gloster TM, Vocadlo DJ. Developing inhibitors of glycan processing enzymes as tools for enabling glycobiology. Nat Chem Biol. 2012;8(8):683-94. DOI: 10.1038/nchembio.1029

[23] Schauer R. Sialic acids as regulators of molecular and cellular interactions. Curr Opin Struct Biol. 2009;19(5):507-14. DOI: 10.1016/j.sbi.2009.06.003

[24] Buschiazzo A, Alzari PM. Structural insights into sialic acid enzymology. Curr Opin Chem Biol. 2008;12(5):565-72. DOI: 10.1016/j.cbpa.2008.06.017

[25] Schauer R. Sialic acids: fascinating sugars in higher animals and man. Zoology (Jena). 2004;107(1):49-64. DOI: 10.1016/j.zool.2003.10.002

[26] Keppler OT, Peter ME, Hinderlich S, Moldenhauer G, Stehling P, Schmitz I, et al. Differential sialylation of cell surface glycoconjugates in a human B lymphoma cell line regulates susceptibility for CD95 (APO-1/Fas)-mediated apoptosis and for infection by a lymphotropic virus. Glycobiology. 1999;9(6):557-69. DOI: 10.1093/glycob/9.6.557

[27] Razi N, Varki A. Cryptic sialic acid binding lectins on human blood leukocytes can be unmasked by sialidase treatment or cellular activation. Glycobiology. 1999;9(11):1225-34.

[28] French BM, Sendil S, Pierson RN, Azimzadeh AM. The role of sialic acids in the immune recognition of xenografts. Xenotransplantation. 2017;24(6). DOI: 10.1111/xen.12345

[29] Wiederschain GYa. Glycobiology: (C. Sansom and O. Markman, eds., Scion Publishing Ltd, UK, 2007, 374 p. DOI: 10.1134/S0006297909010179

[30] Lübbers J, Rodríguez E, van Kooyk Y. Modulation of immune tolerance via Siglec-sialic acid interactions. Front Immunol. 2018;9:2807. DOI: 10.3389/fimmu.2018.02807

[31] Macauley MS, Crocker PR, Paulson JC. Siglec-mediated regulation of immune cell function in disease. Nat Rev Immunol. 2014;14(10):653-66. DOI: 10.1038/nri3737

[32] Zhang C, Chen J, Liu Y, Xu D. Sialic acid metabolism as a potential therapeutic target of atherosclerosis. Lipids Health Dis. 2019;18(1):173. DOI: 10.1186/s12944-019-1113-5

[33] Altman MO, Gagneux P. Absence of Neu5Gc and presence of anti-Neu5Gc antibodies in humans-An evolutionary perspective. Front Immunol. 2019;10:789. DOI: 10.3389/fimmu.2019.00789

[34] Varki NM, Varki A. Diversity in cell surface sialic acid presentations: implications for biology and disease. Lab Invest. 2007;87(9):851-7. DOI: 10.1038/labinvest.3700656

[35] Bhide GP, Colley KJ. Sialylation of N-glycans: mechanism, cellular compartmentalization and function. Histochem Cell Biol. 2017;147(2):149-74. DOI: 10.1007/s00418-016-1520-x

[36] Goswami K, Koner BC. Level of sialic acid residues in platelet proteins in diabetes, aging, and Hodgkin's lymphoma: a potential role of free radicals in desialylation. Biochem Biophys Res Commun. 2002;297(3):502-5. DOI: 10.1016/s0006-291x(02)02241-6

[37] Crook MA. Relationship between plasma sialic acid concentration and microvascular and macrovascular complications in type 1 diabetes. Diabetes care. 2001;24(2):316-22.

[38] Toegel S, Pabst M, Wu SQ, Grass J, Goldring MB, Chiari C, et al. Phenotype-related differential α-2,6- or α-2,3-sialylation of glycoprotein N-glycans in human chondrocytes. Osteoarthritis and Cartilage. 2010;18(2):240-8. DOI: 10.1016/j.joca.2009.09.004

[39] Kuhn B, Benz J, Greif M, Engel AM, Sobek H, Rudolph MG. The structure of human α-2,6-sialyltransferase reveals the binding mode of complex glycans. Acta Crystallogr D Biol Crystallogr. 2013;69(Pt 9):1826-38. DOI: 10.1107/S0907444913015412

[40] Guillot A, Dauchez M, Belloy N, Jonquet J, Duca L, Romier B, et al. Impact of sialic acids on the molecular dynamic of bi-antennary and tri-antennary glycans. Sci Rep. 2016;6(1):35666. DOI: 10.1038/srep35666

[41] Luley-Goedl C, Schmoelzer K, Thomann M, Malik S, Greif M, Ribitsch D, et al. Two N-terminally truncated variants of human β-galactoside α2,6 sialyltransferase I with distinct properties for *in vitro* protein glycosylation. Glycobiology. 2016;26(10):1097-106. DOI: 10.1093/glycob/cww046

[42] Svensson EC, Conley PB, Paulson JC. Regulated expression of alpha 2,6-sialyltransferase by the liver-enriched transcription factors HNF-1, DBP, and LAP. J Biol Chem. 1992;267(5):3466-72.

[43] Appenheimer MM. Biologic contribution of P1 promoter-mediated expression of ST6Gal I sialyltransferase. Glycobiology. 2003;13(8):591-600. DOI: 10.1093/glycob/cwg066

[44] Lee MM, Nasirikenari M, Manhardt CT, Ashline DJ, Hanneman AJ, Reinhold VN, et al. Platelets support extracellular sialylation by supplying the sugar donor substrate. J Biol Chem. 2014;289(13):8742-8. DOI: 10.1074/jbc.C113.546713

[45] Nasirikenari M, Veillon L, Collins CC, Azadi P, Lau JTY. Remodeling of marrow hematopoietic stem and progenitor cells by non-self ST6Gal-1 sialyltransferase. J Biol Chem.

2014;289(10):7178-89. DOI: 10.1074/jbc.M113.508457

[46] Yang WH, Nussbaum C, Grewal PK, Marth JD, Sperandio M. Coordinated roles of ST3Gal-VI and ST3Gal-IV sialyltransferases in the synthesis of selectin ligands. Blood. 2012;120(5):1015-26. DOI: 10.1182/blood-2012-04-424366

[47] Läubli H, Borsig L. Selectins promote tumor metastasis. Semin Cancer Biol. 2010;20(3):169-77. DOI: 10.1016/j.semcancer.2010.04.005

[48] Mócsai A, Walzog B, Lowell CA. Intracellular signalling during neutrophil recruitment. Cardiovasc Res. 2015;107(3):373-85. DOI: 10.1093/cvr/cvv159

[49] Scott DW, Patel RP. Endothelial heterogeneity and adhesion molecules N-glycosylation: Implications in leukocyte trafficking in inflammation. Glycobiology. 2013;23:622-33. DOI: 10.1093/glycob/cwt014

[50] Vestweber D. How leukocytes cross the vascular endothelium. Nat Rev Immunol. 2015;15(11):692-704. DOI: 10.1038/nri3908

[51] de Oliveira S, Rosowski EE, Huttenlocher A. Neutrophil migration in infection and wound repair: going forward in reverse. Nat Rev Immunol. 2016;16(6):378-91. DOI: 10.1038/nri.2016.49

[52] Green CE, Pearson DN, Camphausen RT, Staunton DE, Simon SI. Shear-dependent capping of L-selectin and P-selectin glycoprotein ligand 1 by E-selectin signals activation of high-avidity beta2-integrin on neutrophils. J Immunol. 2004;172(12):7780-90. DOI: 10.4049/jimmunol.172.12.7780

[53] Varki A. Glycan-based interactions involving vertebrate sialic-acid-recognizing proteins. Nature. 2007;446(7139):1023-9. DOI: 10.1038/nature05816

[54] Lin W-L, Guu S-Y, Tsai C-C, Prakash E, Viswaraman M, Chen H-B, et al. Derivation of cinnamon blocks leukocyte attachment by interacting with sialosides. PLoS ONE. 2015;10(6):e0130389. DOI: 10.1371/journal.pone.0130389

[55] Sperandio M. Selectins and glycosyltransferases in leukocyte rolling *in vivo*. FEBS J. 2006;273(19):4377-89. DOI: 10.1111/j.1742-4658.2006.05437.x

[56] Griffin ME, Hsieh-Wilson LC. Glycan engineering for cell and developmental biology. Cell Chem Biol. 2016;23(1):108-21. DOI: 10.1016/j.chembiol.2015.12.007

[57] Miyagi T, Yamaguchi K. Mammalian sialidases: physiological and pathological roles in cellular functions. Glycobiology. 2012;22(7):880-96. DOI: 10.1093/glycob/cws057

[58] Pearce OMT, Läubli H. Sialic acids in cancer biology and immunity. Glycobiology. 2016;26(2):111-28. DOI: 10.1093/glycob/cwv097

[59] Maurice P, Baud S, Bocharova OV, Bocharov EV, Kuznetsov AS, Kawecki C, et al. New insights into molecular organization of human neuraminidase-1: transmembrane topology and dimerization ability. Sci Rep. 2016;6(1):38363. DOI: 10.1038/srep38363

[60] Amith SR, Jayanth P, Franchuk S, Finlay T, Seyrantepe V, Beyaert R, et al. Neu1 desialylation of sialyl α-2,3-linked β-galactosyl residues of TOLL-like receptor 4 is essential for receptor activation and cellular signaling. Cell Signal. 2010;22(2):314-24. DOI: 10.1016/j.cellsig.2009.09.038

[61] Tringali C, Papini N, Fusi P, Croci G, Borsani G, Preti A, et al.

Properties of recombinant human cytosolic sialidase HsNEU2. J Biol Chem. 2004;279(5):3169-79. DOI: 10.1074/jbc.M308381200

[62] Monti E, Bassi MT, Papini N, Riboni M, Manzoni M, Venerando B, et al. Identification and expression of NEU3, a novel human sialidase associated to the plasma membrane. Biochem J. 2000;349(Pt 1):343-51. DOI: 10.1042/0264-6021:3490343

[63] Tanner ME. The enzymes of sialic acid biosynthesis. Bioorg Chem. 2005;33(3):216-28. DOI: 10.1016/j.bioorg.2005.01.005

[64] Varki A. Multiple changes in sialic acid biology during human evolution. Glycoconj J. 2009;26(3):231-45. DOI: 10.1007/s10719-008-9183-z

[65] Bauer J, Osborn HMI. Sialic acids in biological and therapeutic processes: opportunities and challenges. Future Med Chem. 2015;7(16):2285-99. DOI: 10.4155/fmc.15.135

[66] Chiodelli P, Urbinati C, Paiardi G, Monti E, Rusnati M. Sialic acid as a target for the development of novel antiangiogenic strategies. Future Med Chem. 2018;10(24):2835-54. DOI: 10.4155/fmc-2018-0298

[67] Hart GW, Housley MP, Slawson C. Cycling of O-linked beta-N-acetylglucosamine on nucleocytoplasmic proteins. Nature. 2007;446(7139):1017-22. DOI: 10.1038/nature05815

[68] Geisler C, Jarvis DL. Effective glycoanalysis with *Maackia amurensis* lectins requires a clear understanding of their binding specificities. Glycobiology. 2011;21(8):988-93. DOI: 10.1093/glycob/cwr080

[69] Zeng Y, Ramya TNC, Dirksen A, Dawson PE, Paulson JC. High-efficiency labeling of sialylated glycoproteins on living cells. Nat Methods. 2009;6(3):207-9. DOI: 10.1038/nmeth.1305

[70] Du H, Yu H, Ma T, Yang F, Jia L, Zhang C, et al. Analysis of glycosphingolipid glycans by lectin microarrays. Anal Chem. 2019;91(16):10663-71. DOI: 10.1021/acs.analchem.9b01945

[71] Fukasawa T, Asao T, Yamauchi H, Ide M, Tabe Y, Fujii T, et al. Associated expression of α2,3sialylated type 2 chain structures with lymph node metastasis in distal colorectal cancer. Surg Today. 2013;43(2):155-62. DOI: 10.1007/s00595-012-0141-9

[72] Sybirna N, Brodyak I, Zdioruk M, Barska M. Sialic acid-containing glycoproteins are the marker molecules that determine the leukocyte functional state under diabetes mellitus. Sepsis. 2011;4(1):47-55.

[73] Wu G, Nagala M, Crocker PR. Identification of lectin counter-receptors on cell membranes by proximity labeling. Glycobiology 2017;27(9):800-5. DOI: 10.1093/glycob/cwx063

[74] Wang D, Nie H, Ozhegov E, Wang L, Zhou A, Li Y, et al. Globally profiling sialylation status of macrophages upon statin treatment. Glycobiology. 2015;25(9):1007-15. DOI: 10.1093/glycob/cwv038

[75] Antoniuk VO. Lectins and sources of their raw materials. Lviv: Lviv Danylo Halytsky National Medical University; 2005. 554 p. (In Ukrainian).

[76] Ni Y, Tizard I. Lectin-carbohydrate interaction in the immune system. Vet Immunol Immunopathol. 1996;55 (1-3):205-23. DOI: 10.1016/s0165-2427(96)05718-2.

[77] Lee C-Y, Herant M, Heinrich V. Target-specific mechanics of

phagocytosis: protrusive neutrophil response to zymosan differs from the uptake of antibody-tagged pathogens. J Cell Sci. 2011;124(7):1106. DOI: 10.1242/jcs.078592

[78] Rabinovich GA, van Kooyk Y, Cobb BA. Glycobiology of immune responses: Glycobiology of immune responses. Annals of the New York Academy of Sciences. 2012;1253(1):1-15. DOI: 10.1111/j.1749-6632.2012.06492.x

[79] Chiodelli P, Rezzola S, Urbinati C, Federici Signori F, Monti E, Ronca R, et al. Contribution of vascular endothelial growth factor receptor-2 sialylation to the process of angiogenesis. Oncogene. 2017;36(47):6531-41. DOI: 10.1038/onc.2017.243

[80] Sybirna N, Barska M, Brodyak I, Vovk O, Drobot L. A study of carbohydrate determinants of leucocyte glycoprotein receptors in patients with type 1 diabetes mellitus. Annales universitatis Mariae Curie-Sklodovska. 2006;XIX(1,46):215-18.

[81] Ferents I, Brodyak I, Lyuta M, Klymyshyn N, Burda V, Sybirna N. Sialylation status of leukocyte cell-surface glycoconjugates in streptozotocin-induced diabetic rats and after treatment with agmatine. Curr Issues Pharm and Med Sci. 2013;26(4):390-2.

[82] Babu P, North SJ, Jang-Lee J, Chalabi S, Mackerness K, Stowell SR, et al. Structural characterisation of neutrophil glycans by ultra sensitive mass spectrometric glycomics methodology. Glycoconj J. 2009;26(8):975-86. DOI: 10.1007/s10719-008-9146-4

[83] Antonopoulos A, North SJ, Haslam SM, Dell A. Glycosylation of mouse and human immune cells: insights emerging from N-glycomics analyses. Biochem Soc Trans.

2011;39(5):1334-40. DOI: 10.1042/BST0391334

[84] Mondal N, Buffone A, Stolfa G, Antonopoulos A, Lau JTY, Haslam SM, et al. ST3Gal-4 is the primary sialyltransferase regulating the synthesis of E-, P-, and L-selectin ligands on human myeloid leukocytes. Blood. 2015;125(4):687-96. DOI: 10.1182/blood-2014-07-588590

[85] Johnson JL, Jones MB, Ryan SO, Cobb BA. The regulatory power of glycans and their binding partners in immunity. Trends Immunol. 2013;34(6):290-8. DOI: 10.1016/j.it.2013.01.006

[86] Bode L, Rudloff S, Kunz C, Strobel S, Klein N. Human milk oligosaccharides reduce platelet-neutrophil complex formation leading to a decrease in neutrophil β2 integrin expression. J Leukocyte Biology. 2004;76(4):820-6. DOI: 10.1189/jlb.0304198

[87] Caimi G, Montana M, Citarrella R, Porretto F, Catania A, Lo Presti R. Polymorphonuclear leukocyte integrin profile in diabetes mellitus. Clin Hemorheol Microcirc. 2002;27(2):83-9.

[88] Karlsson A. Wheat germ agglutinin induces NADPH-oxidase activity in human neutrophils by interaction with mobilizable receptors. Infect Immun. 1999;67(7):3461-8. DOI: 10.1128/IAI.67.7.3461-3468.1999

[89] Khan F, Khan RH, Sherwani A, Mohmood S, Azfer MA. Lectins as markers for blood grouping. Med Sci Monit. 2002;8(12):RA293-300.

[90] Sybirna N, Zdioruk M, Brodyak I, Bars'ka M, Vovk O. Activation of the phosphatidylinositol-3'-kinase pathway with lectin-induced signal through sialo-containing glycoproteins of leukocyte membranes under type 1

diabetes mellitus. Ukr Biochem J. 2011;83(5):22-31. (In Ukrainian)

[91] Rifat S, Kang TJ, Mann D, Zhang L, Puche AC, Stamatos NM, et al. Expression of sialyltransferase activity on intact human neutrophils. J Leukoc Biol. 2008;84(4):1075-81. DOI: 10.1189/jlb.0706462

[92] Brodyak I, Zdioruk M, Bars'ka M, Vovk O, Sybirna N. The lectincytochemical analyze of mononuclear leucocytes plasmatic membranes sialic-containing glycoprotein of peripheral blood cell under type 1 diabetes mellitus. Med Chem. 2009;4:15-9. (In Ukrainian)

[93] Chen XP, Ding X, Daynes RA. Ganglioside control over IL-4 priming and cytokine production in activated T cells. Cytokine. 2000;12(7):972-85. DOI: 10.1006/cyto.1999.0596

[94] Stamatos NM, Curreli S, Zella D, Cross AS. Desialylation of glycoconjugates on the surface of monocytes activates the extracellular signal-related kinases ERK 1/2 and results in enhanced production of specific cytokines. J Leukoc Biol. 2004;75(2):307-13. DOI: 10.1189/jlb.0503241

[95] Westhorpe CLV, Norman MU, Hall P, Snelgrove SL, Finsterbusch M, Li A, et al. Effector CD4+ T cells recognize intravascular antigen presented by patrolling monocytes. Nat Commun. 2018;9(1):747. DOI: 10.1038/s41467-018-03181-4

[96] Tchilian EZ, Beverley PCL. Altered CD45 expression and disease. Trends Immunol. 2006;27(3):146-53. DOI: 10.1016/j.it.2006.01.001

[97] Amano M, Galvan M, He J, Baum LG. The ST6Gal I sialyltransferase selectively modifies N-glycans on CD45 to negatively regulate galectin-1-induced CD45 clustering, phosphatase modulation, and T cell death. J Biol Chem. 2003;278(9):7469-75. DOI: 10.1074/jbc.M209595200

[98] Swindall AF, Bellis SL. Sialylation of the Fas death receptor by ST6Gal-I provides protection against Fas-mediated apoptosis in colon carcinoma cells. J Biol Chem. 2011;286(26):22982-90. DOI: 10.1074/jbc.M110.211375

[99] Sato T, Furukawa K, Autero M, Gahmberg CG, Kobata A. Structural study of the sugar chains of human leukocyte common antigen CD45. Biochemistry. 1993;32(47):12694-704. DOI: 10.1021/bi00210a019

[100] Zdioruk M, Barska M, Brodyak I, Vovk O, Urbanovich A, Sybirna N. Influence of wortmannin on aggregation ability on neutrophilic granulocytes under type 1 diabetes mellitus. Stud Biol. 2009;3(2):133-40. DOI: 10.30970/sbi.0302.046. (In Ukrainian).

Post-Translational Modifications

Chapter 5

Post-Translational Modifications of Proteins Exacerbate Severe Acute Respiratory Syndrome Coronavirus 2 (SARS-CoV2)

Alok Raghav, Renu Tomar and Jamal Ahmad

Abstract

Severe acute respiratory syndrome coronavirus 2 (SARS-CoV2) is severely affecting the worldwide population. It belongs to the coronavirus family which exhibit protein constituted enveloped single-stranded RNA. These viral proteins undergo post-translational modifications (PTMs) that reorganized covalent bonds and modify the polypeptides and in turn modulate the protein functions. Being viral machinery, it uses host cells system to replicate itself and make their copes, their proteins are also subject to PTMs. Glycosylation, palmitoylation of the spike and envelope proteins, phosphorylation, of the nucleocapsid protein are among the major PTMs responsible for the pathogenesis of the viral infection phase. The current knowledge of CoV proteins PTMs is limited and need to be exploring for to understand the viral pathogenesis mechanism and PTMs effect of infection phase.

Keywords: SARS-CoV-2, Post-translational modifications, Protein, glycosylation palmitoylation, phosphorylation

1. Introduction

SARS-CoV-2 is a culprit of the COVID-19 pandemic, caused 168, 599, 045 infections in people followed by 3,507,477 deaths worldwide as of 28th May 2021 according to the World Health Organization (WHO). Despite protective immunity and vaccination against SARS-CoV-2, it is spreading and affecting the worldwide population causing an increase in severity of illness due to the absence of the pre-existing immunity against SARS-CoV-2. This virus belongs to the family of coronavirus (CoV) with *Nidovirales* order exhibiting protein-containing enveloped positive-strand RNA that causes disease in both humans and animals. SARS-CoV-2 exhibits 30 kb genome size constituted in single-stranded RNA. Its genome comprises 6–11 open reading frames (ORFs) associated with 5′ and 3′ flanking untranslated regions (UTRs).

Coronaviruses are morphologically found spherical accounting average diameter of 80–120 nm with trimeric S-glycoprotein, sometimes with homodimeric HE protein [1, 2]. M-glycoprotein is the most abundant protein of virion which provides structural support to the virion. Moreover, E protein is also an essential protein needed for virion assemble and release [3, 4]. The nucleocapsid of the virus also comprised of the N protein. In the pathophysiology of virus replication, the S protein (180-200 kDa) plays a major role in binding and recognition with the host cell via a cognate receptor(s). This trimeric S protein comprises two S1, S2, HR1 and HR2 subunits. This S protein comprised of two domain on its N-terminal extracellular transmembrane with another short intracellular C-chain domain [5]. The total length of the S protein of SARS-CoV-2 is 1273 amino acids (aa) arranged in a single peptide (1–13 aa) situated at N-terminus, S1 subunit (14–685 aa) and the S2 (686–1273 aa). The last two-sector is responsible for receptor binding and membrane fusion respectively. Furthermore, the S1 subunit comprises of N-terminal domain (14–305 aa) along with the receptor-binding domain (319–541 aa). Similarly, S2 protein constitutes the fusion peptide (788–806 aa), heptapeptide sequence (HR1) (912–984 aa), HR2 (1163–1213 aa), transmembrane domain (1213–1237 aa) along with cytoplasm domain ((1237–1273 aa) [6].

Post-translational modifications (PTMs) refers to the covalent bonds modifications of the proteins post-release from the ribosomes. PTMs adds new functional groups, like phosphate and carbohydrates along with other biological molecules of desired interest. PTMs is a naturally occurring process responsible for the regulation of protein folding, stability, enzymatic activity, protein to protein, and cell-to protein interaction. The most common and routinely occurring PTMs includes glycosylation, phosphorylation and lipidation (addition of such as palmitoylation and myristoylation) via proteolytic cleavage, formation of disulfide bonds. Proteins can also be modified through other covalent modifications like ubiquitination, sumoylation, glycation and neddylation.

PTMs are naturally occurring process catalyzed by enzymes. For instance, N-linked glycosylation needs series of enzyme reaction that generates precursor dolichol-linked oligosaccharide, oligosaccharyltransferase that initiate the transfer of the glycan to a specific consensus sequence (N-X-S/T, where X is any amino acid except proline) along with glycosidases and glycosyltransferases that are needed for the processing of N-linked glycan. In another series of example, protein ubiquitination needed three different enzymes including ubiquitin-activating enzymes (E1), ubiquitin-conjugating enzymes (E2) and ubiquitin ligases (E3) acting sequentially. It is a known fact that a virus utilizes host machinery to replicate its copies, therefore several viral proteins are susceptible to PTMs. Many viral proteins including structural, non-structural and accessory proteins are modified by PTMs which are affecting viral replication and pathogenesis.

2. N-linked glycosylation

N-linked glycosylation of the S protein of coronavirus is reported first time in the 1980s [7]. Murine strain MHV S protein acquire several mannose residues in the rough endoplasmic reticulum (ER). It was demonstrated that the S protein of infectious bronchitis virus (IBV), transmissible gastroenteritis coronavirus (TGEV) and bovine coronavirus is modified by the process of N-liked glycosylation [8]. S protein was also supposed to acquire mannose oligosaccharides and undergoes trimerization post entry to the endoplasmic reticulum (ER) before entering to Golgi complex [9]. In previously published literature the glycosylation sites in S protein of MHV predicted are 20 and 21 in numbers [10].

N-linked glycosylation plays important role in maintaining the conformation of coronavirus S protein, can affect the binding with the receptor of host cells and antigenicity of S protein. In a previously published study of IBV, it was found that mutation on the sites of N-linked glycosylation sites significantly causes shifting in the antigenicity of IBV [11]. In another study, it was found that TGEV infected cells when incubated with tunicamycin (inhibitor of N-linked glycosylation) showed a reduction in the antigenicity of both S and M protein [12]. These research findings suggest that N-linked glycosylation is playing a vital role in maintaining the structure and conformation of the S protein and any mutations occurring at these glycosylation sites contribute to the reduction of antigenicity. It may be proposed here that mutations in the N-linked glycosylation in the S protein of SARS-CoV-2 can be beneficial in controlling the infections and reduction in their antigenicity. In another study conducted on IBV, it was demonstrated that N-D or N-Q mutations at the N-linked glycosylation sites N212 or N276 inhibit the function of S protein and in turn, hampers the cell-to-cell fusion and recognition [13].

Moreover, N-linked glycosylation of Dipeptidyl-peptidase 4 (DPP4), a similar receptor of Middle East respiratory syndrome coronavirus (MERS-CoV) significantly affect the binding of MERS-CoV S protein. E protein of SARS-CoV is another important constituent of the virus, it has been demonstrated that it contains two N-linked glycosylation sites N48 and N66, while IBV E protein contains one site at N5. In one of the studies published in past associated with SARS-CoV, it was demonstrated that transfected E protein with N-terminal Myc-tag showed glycosylation co-translationally [14]. However, two transmembrane domains are required to interact with SARS-CoV M protein and the hydrophilic region [14].

In previously published literature it was demonstrated that M protein of α-coronavirus transmissible gastroenteritis virus (TGEV) and Porcine Epidemic Diarrhea Virus (PDEV), gamma coronavirus IBV along with turkey enteric coronavirus are susceptible to N-linked glycosylation which can be inhibited by the tunicamycin [15–17]. The SARS-CoV-2 S glycoprotein possesses 22 N-linked glycosylation sites as confirmed from the recently published literature [18–20]. SARS-CoV-2 S glycoprotein showed a conserved S2 subunit for N-linked glycosylation with a low tendency for O-linked glycosylation. The N-linked glycosylation in SARS-CoV-2 is featured by binding of GlcNAc with the Asp amino acid residue in the Asp-X-Ser/Thr consensus sequence in which the residue X is amino acid except for proline.

It is known that N-linked glycosylation is compulsory for understanding the location, structure and infectivity of the viruses and also their interaction with the host cells. These glycoproteins plays important role in immune responses but still need more research [21, 22]. The S glycoprotein represents unique pathogen-associated molecular patterns (PAMPs) that further recognized by the host pattern Recognition Receptors (PPRs). These PPRs includes Toll-like receptors 3, 4, 7, 8 and 9 along with C-type lectins and collectins [23, 24]. SARS-CoV is recognized by the toll-like receptors 3 and 4 via MyD88 and TRIF, furthermore, the same process may be proposed for the pathogenesis and infectivity in SARS-CoV2 [22, 25].

3. O-linked glycosylation

O-linked glycosylation involved in providing structural and functional stability to protein and believed to play important role in the maintenance of viral entity and biological activities associated with these viral proteins [26]. It has been demonstrated from a previously published study that Ser673, Thr678 and Ser686 are the conserved sites of O-linked glycosylation in human SARS-CoV-2 and other coronaviruses especially in S protein [26]. Moreover, O-glycosylation sites were predicted using the tool Net-O-Gly server 4.0 and found three sites for O-linked glycosylation at Ser673, Thr678 and Ser686 [26]. In another study, it was found that O-glycosylation at Thr 323 and Ser 325 and Thr 323 of the S1 glycoprotein are the possible and predicted sites of O-linked glycosylation in SARS-CoV-2 viral proteins [27].

The O-linked glycosylation at the Thr323 is confirmed by the presence of proline amino acids at position 322, making the possibility that the presence of proline amino acid is higher adjacent to the O-linked glycosylation sites [28]. Cryo-electron microscopic images of the SAS-CoV2 indicate that the binding of S protein to the human angiotensin I-converting enzyme 2 (hACE2) receptor involves an association between receptor-binding domain (RBD) and the hACE2 peptidase domain [29, 30]. The RBD of the S protein in the S1 subunit endures hinge-like dynamic movement of accelerating the detention of RBD with hACE2, exhibiting a 10–20 fold increase in affinity for the hACE2 receptors [31, 32].

In another published study it was found that application of tunicamycin showed normal glycosylation of the M protein despite inhibiting the N-linked glycosylation of S protein [33]. In one of the study, the structures of the associated glycans to the M protein during the O-linked glycosylation showed that it added into two-step processes; GalNAc first added before the addition of galactose and the sialic acid. After the possession of the GalNAc, galactose and sialic acid sequentially, the M protein was further undergone modification in the trans-Golgi apparatus [34].

Recently it was reported that there are low levels of O-linked glycosylation in the S protein of SARS-CoV2 [35]. These glycans regulate the recognition of the antibodies and impinge on priming by the host proteases enzyme system. Mucin-type O-linked glycosylation is featured with the presence of GalNAc associated with the hydroxyl group of serine and threonine amino acid residues. Mucins contain a significant number of O-GalNAc glycans [36]. The presence of the O-linked glycans involved in the O-linked glycosylation of viral proteins suggests a vital role in biological activity. In the SARS-CoV-2 S1 protein, the O-linked glycosylation as GalNAc and O-GlcNAc appears to be involved in the structural and functional stability of the protein. The current scenario involves the use of a vaccine that utilizes the S protein glycosylation as a target.

4. Palmitoylation

Palmitoylation refers to the attachment of the palmitic fatty acid to the cysteine (S-palmitoylation) but less frequently to the serine and threonine amino acid residues (O-palmitoylation). In coronaviruses studies, the S protein undergone palmitoylation in infected cells and in the presence of tunicamycin it does not undergoes palmitoylation [37]. In another study conducted on MHV S protein reduces infectivity of MHV when treated with palmitoyl acyltransferase inhibitors 2-bromopalmitate [38]. The cytoplasmic part of the SARS-CoV S protein consists of four cysteine-rich clusters among them 2 clusters modified upon palmitoylation. However, cell surface expression of SARS-CoV S protein was unaffected due to this palmitoylation. In one of the previously published study, it was found that treatment of nitric oxide significantly leads to a reduction in the palmitoylation of the S protein of SARS-CoV [39].

In one of the study, it was found that there is three cysteines at position 40, 43 and 44 are found to undergo palmitoylation in the E protein of the SARS-CoV [40]. In another study, homologous cysteine of the E protein of MV-A59 at position C40, C44 and C47 were mutated to the alanine residues as resultant infectivity decreased [41]. It is therefore concluded that palmitoylation of the MHV E protein contributes to the stability and biological activity of the mature virions. Contrary, palmitoylation of the SARS CoV E protein is not mandatory for its interception with N protein.

5. Phosphorylation

In SARS-CoV2 the most abundant genomic protein encoded is the N protein with a significantly higher level of translation at the early stage of the infection. In all form of the coronaviruses, the N-protein is almost the same and conserved containing two globular domains, the N- terminal domain (NTD) and the C-terminal domain (CTD). Around these domains, intrinsically disorganized regions are present. N protein is dimeric with multiple RNA binding sites including one major RNA-binding groove, which is created by the two CTD piling on each other on NTD [42]. In previously published literature it has been found that in the disorganized region, there is an abundance of serine-arginine residues that is essential for the essential function and regulation of the N-protein [42]. Cytoplasmic kinases mediate the phosphorylation of the N-protein in the early infection phase. N protein, a helical nucleocapsid, constituent of SARS-CoV2 virus arranged in beads on a string pattern which shows binding with the RNA. It has two domains namely the N-terminal domain and C-terminal domain. It has been found that both domains contribute to the binding of the viral genome. In one of the recently published study, it was demonstrated in truncation analysis that an L/Q-rich region placed within the intrinsically disordered region of the SARS-CoV-2 N protein plays a vital role in RNA-mediated phase separation, which is located adjacent to the phosphorylated SR-rich region (constituting residues 176–206) [43]. In the same study, it was concluded that N protein central intrinsically disordered region shown to be involved in protein–protein interactions mediated via putative hydrophobic α-helix spanning residues (213–225 residues) [43].

6. Prospects

The functional role of PTMs in SARS-CoV2 associated proteins has not been fully explored and many trials are needed for proposing the role of all types of glycosylation like N-linked and O-linked along with palmitoylation and phosphorylation in the initial phase of the infection. However multiple modification sites on the proteins of the SARS-CoV2 virus provides opportunities to explore more about the replication and pathogenesis of the virus into the host cells. Moreover, newer techniques for the detection of the PTMs are also needed to detect the modifications at multiple sites in dynamically changing virus structure. It is also needed in the current scenario to better understand the molecular mechanism of these PTMs. Also, the PTMs of the coronavirus proteins might be attractive targets for the therapeutic regime. PTMs of the coronavirus proteins might also provide a prospective target for the development of the vaccines.

Author details

Alok Raghav[1*], Renu Tomar[2] and Jamal Ahmad[3]

1 Multidisciplinary Research Unit, GSVM Medical College, Kanpur, India

2 Department of Public Health, Poornima University, Jaipur, India

3 Rajiv Gandhi Centre for Diabetes and Endocrinology, Aligarh Muslim University, Aligarh, India

*Address all correspondence to: alokalig@gmail.com

IntechOpen

References

[1] Masters PS. The molecular biology of coronaviruses. Advances in Virus Research. 2006;**66**:193-292

[2] de Groot RJ. Structure, function and evolution of the hemagglutinin-esterase proteins of corona-and toroviruses. Glycoconjugate Journal. 2006;**23**(1-2): 59-72

[3] Liu DX, Inglis SC. Association of the infectious bronchitis virus 3c protein with the virion envelope. Virology. 1991;**185**(2):911-917

[4] To J, Surya W, Fung TS, et al. Channel-inactivating mutations and their revertant mutants in the envelope protein of infectious bronchitis virus. Journal of Virology. 2017;**91**(5): e02158-e02116

[5] Huang Y, Yang C. Xu, Xf. et al. Structural and functional properties of SARS-CoV-2 spike protein: potential antivirus drug development for COVID-19. ActaPharmacol Sin. 2020;**41**: 1141-1149

[6] Xia S, Zhu Y, Liu M, Lan Q, Xu W, Wu Y, et al. Fusion mechanism of 2019-nCoV and fusion inhibitors targeting HR1 domain in spike protein. Cellular & Molecular Immunology. 2020;**17**:765-767

[7] Niemann H, Boschek B, Evans D, Rosing M, Tamura T, Klenk HD. Post-translational glycosylation of coronavirus glycoprotein E1: inhibition by monensin. The EMBO Journal. 1982;**1**(12):1499-1504

[8] Delmas B, Laude H. Assembly of coronavirus spike protein into trimers and its role in epitope expression. Journal of Virology. 1990;**64**(11): 5367-5375

[9] Nal B, Chan C, Kien F, et al. Differential maturation and subcellular localization of severe acute respiratory syndrome coronavirus surface proteins S, M and E. The Journal of General Virology. 2005;**86**(Pt 5):1423-1434

[10] Schmidt I, Skinner M, Siddell S. Nucleotide sequence of the gene encoding the surface projection glycoprotein of coronavirus MHV-JHM. The Journal of General Virology. 2010;**68**(Pt 1):47-56

[11] Smati R, Silim A, Guertin C, et al. Molecular characterization of three new avian infectious bronchitis virus (IBV) strains isolated in Quebec. Virus Genes. 2002;**25**(1):85-93

[12] Delmas B, Laude H. Carbohydrate-induced conformational changes strongly modulate the antigenicity of coronavirus TGEV glycoproteins S and M. Virus Research. 1991;**20**(2): 107-120

[13] Zheng J, Yamada Y, Fung TS, Huang M, Chia R, Liu DX. Identification of N-linked glycosylation sites in the spike protein and their functional impact on the replication and infectivity of coronavirus infectious bronchitis virus in cell culture. Virology. 2018;**513**:65-74

[14] McBride CE, Machamer CE. Palmitoylation of SARS-CoV S protein is necessary for partitioning into detergent-resistant membranes and cell-cell fusion but not interaction with M protein. Virology. 2010;**405**(1): 139-148

[15] Hogue BG, Nayak DP. Expression of the porcine transmissible gastroenteritis coronavirus M protein. Advances in Experimental Medicine and Biology. 1990;**276**:121-126

[16] Utiger A, Tobler K, Bridgen A, Ackermann M. Identification of the membrane protein of porcine epidemic

diarrhea virus. Virus Genes. 1995;**10**(2): 137-148

[17] Dea S, Verbeek AJ, Tijssen P. Antigenic and genomic relationships among turkey and bovine enteric coronaviruses. Journal of Virology. 1990;**64**(6):3112-3118

[18] Wang Q, Zhang Y, Wu L, et al. Structural and functional basis of SARS-CoV-2 entry using human ACE2. Cell. 2020;**181**:894-904.e9

[19] Wu D, Wu T, Liu Q, et al. The SARS-CoV-2 outbreak: what we know. International Journal of Infectious Diseases. 2020;**94**:44-48

[20] Yuan M, Wu NC, Zhu X, et al. A highly conserved cryptic epitope in the receptor-binding domains of SARS-CoV-2 and SARS-CoV. Science. 2020;**368**:630-633

[21] Hompson AJ, de Vries RP, Paulson JC. Virus recognition of glycans receptors. CurrOpinVirol. 2019;**34**: 117-129

[22] van Kooyk Y, Rabinovich GA. Protein-glycan interactions in the control of innate and adaptive immune responses. Nature Immunology. 2008;**9**:593-601

[23] Kawai T, Akira S. The role of pattern-recognition receptor in innate immunity: update on Toll-like receptor. Nature Immunology. 2010;**11**:373-384

[24] Alexopoulou L, Holt AC, Medzhitov R, et al. Recognition of double-stranded RNA and activation of NF-kappaB by toll-like receptor 3. Nature. 2001;**413**:732-738

[25] Li G, Fan Y, Lai Y, et al. Coronavirus infections and immune responses. Journal of Medical Virology. 2020;**92**:424-432

[26] Andersen KG, Rambaut A, Lipkin WI, Holmes EC, Garry RF. The proximal origin of SARS-CoV-2. Nature Medicine. 2020;**26**:450-452

[27] Uslupehlivan M, Sener E. Glycoinformatics approach for identifying target positions to inhibit initial binding of SARS-CoV-2 S1 protein to the host cell. bioRxiv. 2020. DOI: 10.1101/2020.03.25.007898

[28] ThankaChristlet TH, Veluraja K. Database analysis of O-glycosylation sites in proteins. Biophysical Journal. 2001;**80**:952-960

[29] Walls AC, Park YJ, Tortorici MA, Wall A, McGuire AT, Veesler D. Structure, function, and antigenicity of the SARS-CoV-2 spike glycoprotein. Cell. 2020;**180**:1-12

[30] Hoffmann M, Kleine-Weber H, Schroeder S, Kruger N, Herrler T, Erichsen S, et al. SARS-CoV-2 cell entry depends on ACE2 and TMPRSS2 and is blocked by a clinically proven protease inhibitor. Cell. 2020;**181**:1-10

[31] Wrapp D, Wang N, Corbett KS, Goldsmith JA, Hsieh CL, Abiona O, et al. Cryo-EM structure of the 2019-nCoV spike in the prefusion conformation. Science. 2020;**367**: 1260-1263

[32] Yan R, Zhang Y, Li Y, Xia L, Guo Y, Zhou Q. Structural basis for the recognition of SARS-CoV-2 by full-length human ACE2. Science. 2020;**367**:1444-1448

[33] Holmes KV, Doller EW, Sturman LS. Tunicamycin resistant glycosylation of coronavirus glycoprotein: demonstration of a novel type of viral glycoprotein. Virology. 1981;**115**(2): 334-344

[34] Locker JK, Griffiths G, Horzinek MC, Rottier PJ. O-glycosylation of the coronavirus M protein. Differential localization of sialyltransferases in N- and O-linked

glycosylation. The Journal of Biological Chemistry. 1992;**267**(20):14094-14101

[35] Zhao P, Praissman JL, Grant OC, et al. Virus-receptor interactions of glycosylated SARS-CoV-2 spike and human ACE2 receptor. Cell Host & Microbe. 2020;**24**:S1931-3128(20) 30457-1

[36] Bagdonaite I, Wandall HH. Global aspect of viral glycosylation. Glycobiology. 2018;**28**:443-467

[37] van Berlo MF, van den Brink WJ, Horzinek MC, van der Zeijst BA. Fatty acid acylation of viral proteins in murine hepatitis virus-infected cells. Brief Report. Archives of Virology. 1987;**95**(1-2):123-128

[38] Thorp EB, Boscarino JA, Logan HL, Goletz JT, Gallagher TM. Palmitoylations on murine coronavirus spike proteins are essential for virion assembly and infectivity. Journal of Virology. 2006;**80**(3):1280-1289

[39] Akerstrom S, Gunalan V, Keng CT, Tan Y-J, Mirazimi A. Dual effect of nitric oxide on SARS-CoV replication: viral RNA production ¨and palmitoylation of the S protein are affected. Virology. 2009;**395**(1):1-9

[40] Liao Y, Yuan Q, Torres J, Tam JP, Liu DX. Biochemical and functional characterization of the membrane association and membrane permeabilizing activity of the severe acute respiratory syndrome coronavirus envelope protein. Virology. 2006;**349**(2):264-275

[41] Boscarino JA, Logan HL, Lacny JJ, Gallagher TM. Envelope protein palmitoylations are crucial for murine coronavirus assembly. Journal of Virology. 2008;**82**(6):2989-2999

[42] Carlson CR et al. Phosphorylation Modulates Liquid-Liquid Phase Separation of The Sars-Cov-2 N Protein.

bioRxiv preprint2020. DOI: 10.1101/ 2020.06.28.176248

[43] Lu S, Ye Q, Singh D, et al. The SARS-CoV-2 nucleocapsid phosphoprotein forms mutually exclusive condensates with RNA and the membrane-associated M protein. Nature Communications. 2021;**12**:502

www.ingramcontent.com/pod-product-compliance
Lightning Source LLC
Chambersburg PA
CBHW081233190326
41458CB00016B/5770